front cover bees at honeycomb
R. Morley/Natural Science Photos
back cover water flea with eggs inside
brood pouch   Heather Angel

Purnell's Concise
Dictionary of Nature

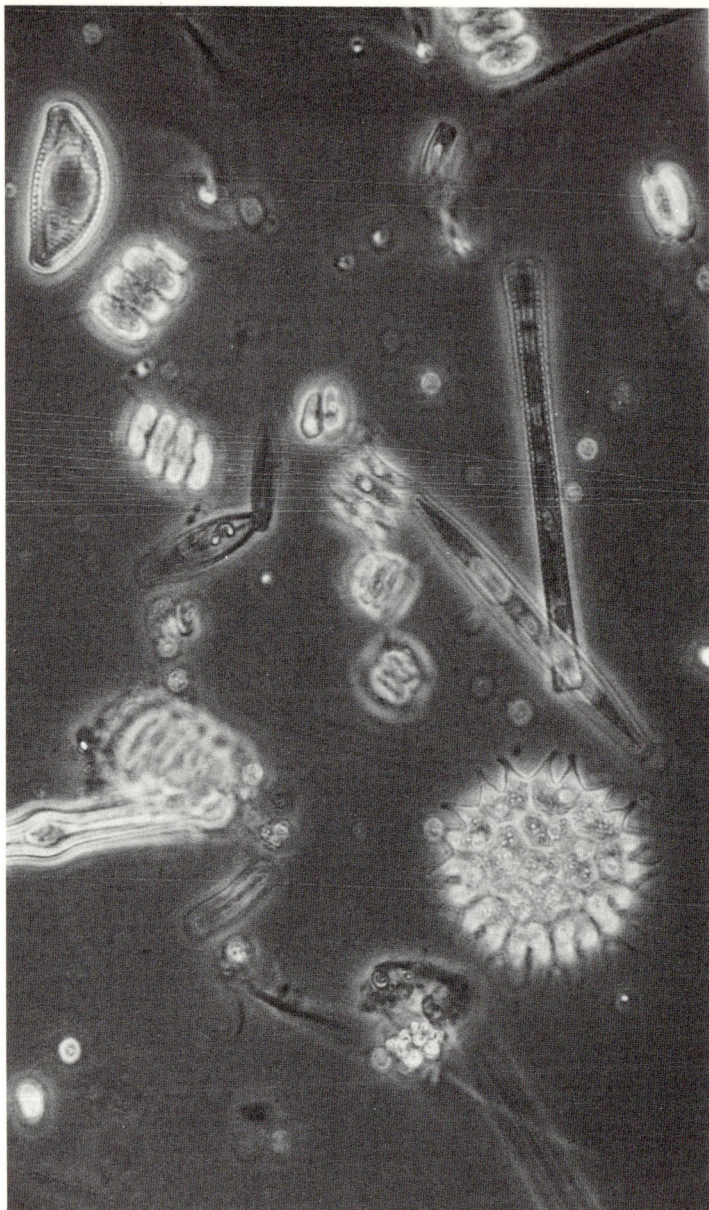

Green algae

Purnell's Concise
Dictionary
of Nature

Michael Dempsey

Illustrated by
David Wright

PURNELL

ISBN 0 361 06293 1
Copyright © 1984 Purnell Publishers Limited
Published 1984 by Purnell Books, Paulton, Bristol BS 18 5LQ
a member of the BPCC group of companies
Made and printed in Great Britain by
Hazell Watson & Viney Ltd, Aylesbury, Bucks

A

Aardvark

Aardvark A strange African mammal with no close relatives. It has a bulky grey body about one metre in length, a long narrow head and large ears. The aardvark feeds mainly on termites, which it digs out of the ground with its powerful claws and laps up with a long sticky tongue.

Aardwolf An African member of the hyena family, somewhat larger than a fox. The coat is yellowish grey with black stripes. It is a nocturnal animal and feeds mainly on termites, which it picks up with its long tongue.

Abalone A single-shelled marine mollusc related to limpets and also known as an ormer or earshell. The body is snail-like and fringed with tentacles. The shell has a line of holes across it, through which water is exhaled. Some abalones are among the largest of shellfish. They live in many parts of the world but are commonest in warmer waters.

Abcission layer A layer at the base of the petiole of a woody plant that breaks down and causes leaf fall in the autumn.

Abyssinian cat, *see* CAT

Acacia A large group of EVERGREEN trees and shrubs, found mainly in the warm and dry parts of the world, including the African savanna and the Australian bush country, where they are called wattles. Many are thorny and bear cylindrical or globular clusters of yellow or white flowers. Some are known as mimosas.

Acanthus Any of a group of plants with spikes of white or purple flowers and large toothed or spiky leaves. They are found in Africa, Asia and Europe, and like dry climates.

Accentor A member of a small family of birds found in Europe and Asia. The commonest species is the dunnock, also called the hedge sparrow, although it is not a sparrow at all. Accentors have slender beaks and feed on insects and small seeds.

Achene A type of fruit which normally contains only one seed and which is indehiscent — i.e. it does not split open and release the seed. The fruit wall is generally leathery: if woody, the fruit is called a nut.

Acorn worm A strange animal that lives in the mud of the seabed, often at great depths. The 'acorn' is the head end of the animal, which sticks out of the mud and is bright orange or red. There are several species, ranging from about 5 cm to 2 metres in length. The body is rather like that of an earthworm at first glance, but the animal breathes by means of gills. It is something of a link between the invertebrates and the vertebrates.

Actinomorphic flowers are regularly arranged and can be cut into two similar halves in two or more directions.

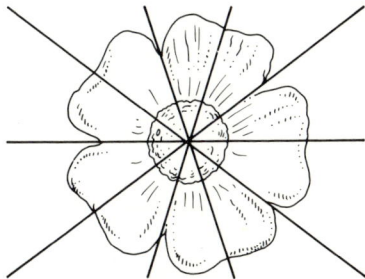

Actinomorphic flower

Actomyosin A fibrous compound of two proteins, actin and myosin, present in muscles. Shortening of the fibres causes muscular contraction and enables movement.

Adaptation A change in an animal's body to counteract a change in the environment. For example, some mammals adapt to warmer summer temperatures by moulting. The word adaptation is also used in connection with evolution to mean a characteristic that enables an animal to survive under certain conditions.

Addax A rare, whitish antelope, related to the oryx. It lives in the deserts of North Africa and can live almost indefinitely without drinking. It gets enough water from the plants that it eats and from the dew on them.

Adder A snake of the viper family, often simply called the viper. It has a fairly stout body up to about 60 cm long, and a characteristic black zigzag mark along the back. It lives in Asia and Europe and it is found further north than other snakes. It is the only poisonous snake in Britain. It feeds mainly on lizards, mice and voles.

Adélie penguin A small penguin standing about 45 cm high which lives on Antarctica and neighbouring islands. It has simple black and white markings and a comic, waddling gait. To move faster, these penguins toboggan over the ice on their bellies, using their feet and flippers to propel themselves.

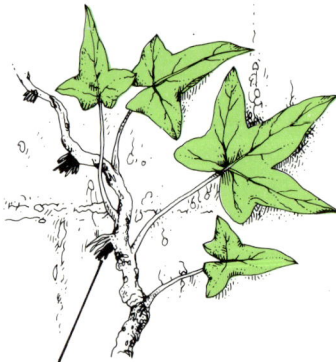

Adventitious roots of ivy

Adenoids, *see* TONSILS

Adrenal gland These tiny glands, found in all mammals, are loosely attached to the kidneys. They play an important part in the regulation of the body's activity by releasing hormones, such as ADRENALINE, in response to stress or tension.

Adrenaline A hormone secreted by the adrenal gland in times of fear or anxiety. By increasing the heart-rate (so creating the thumping of the heart) and the amount of blood flowing to brain and muscles, it increases the body's efficiency in readiness to face threatened danger.

Adventitious roots Roots which develop from the stem or leaf, rather than as branches from the main root.

Aerobic Requiring free oxygen for respiration.

Aestivation Animals in many of the drier parts of the world hide away in the summer or driest season and sleep, thereby avoiding the danger of drying up. This is known as aestivation.

African violet Tropical African flower often grown as a house plant. It has furry heart-shaped leaves, and is not a true violet.

Agama A group of lizards related to the iguanas. The common agama of West Africa is about 30 cm long and has a variety of colours ranging from brown to vivid orange and blue. Other agamas include the MOLOCH and FRILLED LIZARD of Australia.

Agaric The agarics are a large group of fleshy fungi with gills. They include many familiar fungi such as the field mushroom, the chanterelle, the red and white spotted fly agaric and the notorious death cap.

Agave A group of desert-living American plants with rosettes of thick spiky leaves up to 3 metres long. Dense clusters of pale flowers are carried on tall central stalks. Some species take many years to flower and are called century plants. Sisal and other fibres are obtained from the leaves.

Agouti A South American rodent looking like a long-legged guinea pig. There are several different kinds, all brownish in colour and about 50 cm long. Agoutis live mainly in woods and feed on leaves, roots and fallen fruit.

Ailanthus, *see* TREE OF HEAVEN

Albatross A group of birds related to the petrel, and characterized by tube-like nostrils on top of the beak. Albatrosses are large birds, with goose-sized bodies and very long, narrow wings. The wandering albatross has a wing span of 3.5 metres or more — greater than that of any other bird. Most live in the southern hemisphere and they spend nearly all their lives at sea, gliding gracefully over the waves.

Albinism Lack of pigment in the skin, resulting in pure white individuals. Albino mammals and other vertebrates have pink eyes because there is no pigment in the iris to mask the colour of the blood capillaries.

Fly agaric

Alder A deciduous tree or shrub of Europe, Asia and the Americas. The male flowers are carried in swaying catkins and the female flowers in small cone-like catkins, all opening before the leaves. Most kinds grow near water.

Alder fly An insect related to the lacewings. The young alder fly lives in water and the adults are never found far from ponds or streams. They have rather smoky wings with thick, black veins. Fishermen often use artificial alder flies as bait for trout.

Alevin, *see* SALMON

Alewife, *see* SHAD

Alfalfa A hardy crop plant, also known as lucerne, which is widely grown as animal feed. Like other members of

the pea family, it restores nitrogen to the soil by way of its root nodules. Each leaf has three leaflets on a short stalk.

Algae (singular alga) A large division of flowerless plants, most of which live in water. They include many single-celled planktonic organisms of fresh and salt water, the filamentous pond scums, and the various-coloured seaweeds.

Alimentary canal The food canal or gut of an animal running from the mouth to the anus, in which DIGESTION takes place and from which simple food compounds are absorbed into the rest of the body.

Alligator A large reptile closely related to the crocodiles, but distinguished by its blunter snout and by the fact that the fourth tooth in the lower jaw is not visible when the mouth is shut. Alligators live in warm rivers and spend much of their time basking on the banks. Adults feed on fish, birds, and small mammals that come to the river to drink or swim. There are two species of alligator, one in North America reaching 4.5 metres or more in length, and a much smaller one in China.

Alligator-pear, *see* AVOCADO

Alligator snapping turtle, *see* SNAPPING TURTLE

Allis shad, *see* SHAD

Allspice The dried berries of the West Indian pimento; it tastes like a blend of cinnamon, clove and nutmeg.

Almond A tree bearing pink flowers and leathery green stone fruits. The stones, commonly but mistakenly called nuts, contain edible kernels. Sweet kernels are eaten; bitter ones used for oil. The tree grows best in warm regions.

Aloe A group of perennial plants of southern Africa with stiff fleshy leaves and red or yellow flowers. Aloes range from a few centimetres to 30 metres tall. The leaves yield a bitter juice which is used for medicinal purposes and fibre which is used for rope and cloth.

Alpaca A domesticated form of the guanaco with a long soft coat, normally either totally black or white. Alpacas are found in Bolivia and Peru. Their wool is clipped and made into expensive cloth.

Alpine marmot

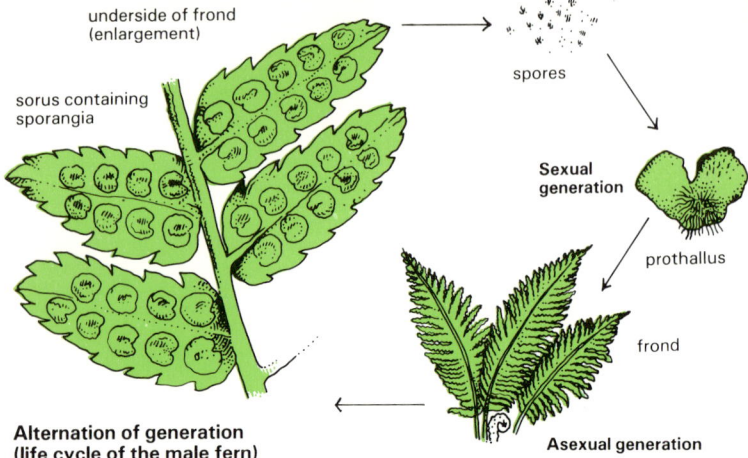

underside of frond (enlargement)

spores

sorus containing sporangia

Sexual generation

prothallus

frond

Alternation of generation (life cycle of the male fern)

Asexual generation

Alpine marmot, *see* MARMOT

Alternation of generations The existence of two distinct forms in the life-cycle of an organism, both of which can reproduce. One form reproduces sexually and gives rise to the other form which in turn reproduces asexually to give the sexual form again. This phenomenon occurs in many plants, notably the FERNS, but is less common in animals.

Alveolus (plural alveoli) The name given to the millions of minute air-sacs that make up the lung. It is across the thin walls of the alveoli that the exchange of oxygen and carbon dioxide occurs, between blood and air, during respiration. The term is also used to describe the socket into which teeth fit.

Amaranth A family of warm-climate plants whose flowers keep their colour even when dried. Love-lies-bleeding is a popular species.

Amaryllis The belladonnna lily, a bulbous plant from southern Africa which is cultivated for its red fragrant flowers. The name amaryllis is also used by horticulturists for related plants whose thick stalks bear showy clusters of lily-like flowers. The amaryllis has given its name to the narcissus and snowdrop family, the Amaryllidaceae.

Amino-acid An organic substance containing carbon, hydrogen, oxygen and nitrogen (and in a few cases sulphur). Various combinations of the 25 or so amino-acids make up all the known proteins.

Amnion The embryos of reptiles, birds and mammals grow in a sac of liquid. The wall of this sac is the amnion.

Amoeba A protozoan which usually lives in water. The animal is continuously changing its shape as it moves about. It feeds by engulfing tiny bacteria and other particles. There are many kinds, the largest being just visible to the naked eye.

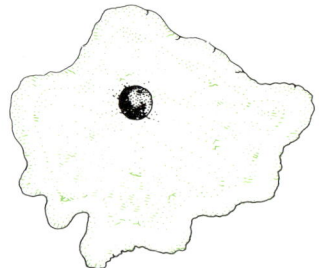

Amoeba

Amphibian A cold-blooded vertebrate belonging to the class Amphibia that usually needs to return to the water to breed and, except in some limbless forms, lacks scales. Amphibians undergo METAMORPHOSIS during development, though in some cases, such as the AXOLOTL, the process is incomplete. Adults usually have lungs and a moist skin through which they can breathe. Examples include FROGS, TOADS, NEWTS and SALAMANDERS.

Amphipod A group of small shrimp-like marine and freshwater crustaceans. Among the most familiar are the sand-hoppers which live among seaweed along the strand line and jump about when disturbed.

Amphisbaenid A group of worm-like burrowing lizards found in the warmer parts of Europe, Africa and the Americas. With legless bodies and scales arranged in rings, they closely resemble large earthworms.

Amphiuma A strange amphibian of the south-eastern USA with a snake-like body up to one metre long and tiny legs. Amphiumas spend much of their time in underwater burrows, raising their heads above the water to breathe.

Anabolism The building up of complex molecules from simpler ones in the body of a plant or animal.

Anaconda The largest snake in the world, with lengths often exceeding 7 metres. The anaconda lives in South America and spends a lot of time in the water. It is often called the water boa. It feeds mainly on birds, deer and rodents. It is not a poisonous snake, but kills by twining around and suffocating its prey.

Anaerobic Able to survive in the absence of free oxygen.

Anatomy The study of the structure, both internal and external, of animals.

Anchovy A small herring-like fish, of which there are several species. Anchovies are most abundant in warmer seas, but they are also found in the shallower regions and bays of cooler seas. Like the herring, the anchovy lives in huge shoals and feeds on the tiny plankton that floats near the sea surface.

Androecium The male parts (i.e. the stamens) of a flower.

Anemone A spring flower, often called windflower, growing in northern woods. Anemones have pale, delicate blooms, but cultivated ones are bigger and more brightly coloured.

Angelfish The name commonly given to three quite unrelated fishes with wing-like fins and bright colours. One is a shark (the MONKFISH), one a small fish of coral reefs, and one a freshwater fish which is a great favourite for aquaria.

Angelfish

Anglerfish A group of marine fishes which catch their food with a sort of 'fishing rod'. The 'rod' is a modified fin spine, and it hangs near the mouth with a worm-like flap of skin acting as the bait. Small fishes investigating the bait are quickly snapped up. The anglerfishes have very large heads and mouths. They are found in warm and temperate seas and most live on the bottom. Some species live at great depths.

Anglerfish

13

Angwantibo A rare member of the loris family found in the forests of West Africa. The angwantibo has a furry body about 25 cm long, large eyes and relatively long limbs. There is no visible tail. The hands and feet are highly developed for grasping branches and angwantibos move nimbly through the trees.

Ani The anis are insect-eating birds of North and South America. They are related to the cuckoos, although they do not lay their eggs in other birds' nests. Their nesting behaviour is unusual, however, because they live in colonies and several pairs usually combine to build a communal nest. The birds are about 40 cm long and shiny black in colour.

Animal kingdom The chart on the following pages depicts the major groups of animals in the form of a family tree, showing the evolutionary relationship between the groups.

Anise An annual herb from the Mediterranean area belonging to the carrot family. It is grown for its spicy seeds which are used to flavour sweet foods.

Annelid A member of the phylum Annelida which contains the segmented worms — earthworms, bristleworms, leeches and others — whose bodies are normally clearly divided into segments or rings.

Annual A plant that completes its entire life cycle, from seed to seed, and dies in a single season.

Annual ring The increase in girth shown each year in a woody stem. Because, in temperate climates, there is a distinct difference between XYLEM formed in spring and in autumn, the rings give the approximate age of the tree. In tropical climates, however, growth is constant throughout the year and no rings are seen. In temperate climates it is possible for more than one growth ring to form in one year. This may happen when a very cold spell occurs during the summer.

Anoa Dwarf cattle standing one metre high at the shoulder and resembling antelopes in general appearance. Anoas are found in the mountains of the Celebes islands in Indonesia.

Anole A group of small slender lizards with long tails and triangular heads. They are related to the iguanas and are found only in the Americas. Most of them live in the trees and climb by means of sharp claws and adhesive pads on the toes. Anoles can change colour with their mood, from brown to green.

Ant A large group of insects belonging to the order Hymenoptera, which also contains the bees and wasps. There are several thousand kinds of ant and they can usually be recognized by their slender 'waists' and sharply bent antennae. The ants all live in communities, and most of the work is done by wingless worker ants. Winged ants occur at certain times of the year and fly off to mate. The males then die and the females or queens form their own nests or swell the size of existing ones.

Antarctic cod A common fish of antarctic waters named for its similarity to cod in taste. It has a large head and a small, flattened body.

Antbird A family of small birds found in the forests of Central and South America. Despite their name, antbirds do not feed chiefly on ants; larger insects, snails and worms are more typical food items.

Anteater The three kinds of anteater live in South and Central America and they feed almost exclusively on ants and termites. The giant anteater has a remarkably long cylindrical snout and a huge bushy tail. The whole animal is 2 metres long. The silky anteater and the two-toed anteater (or tamandua) are smaller and they live in the trees.

Angwantibo

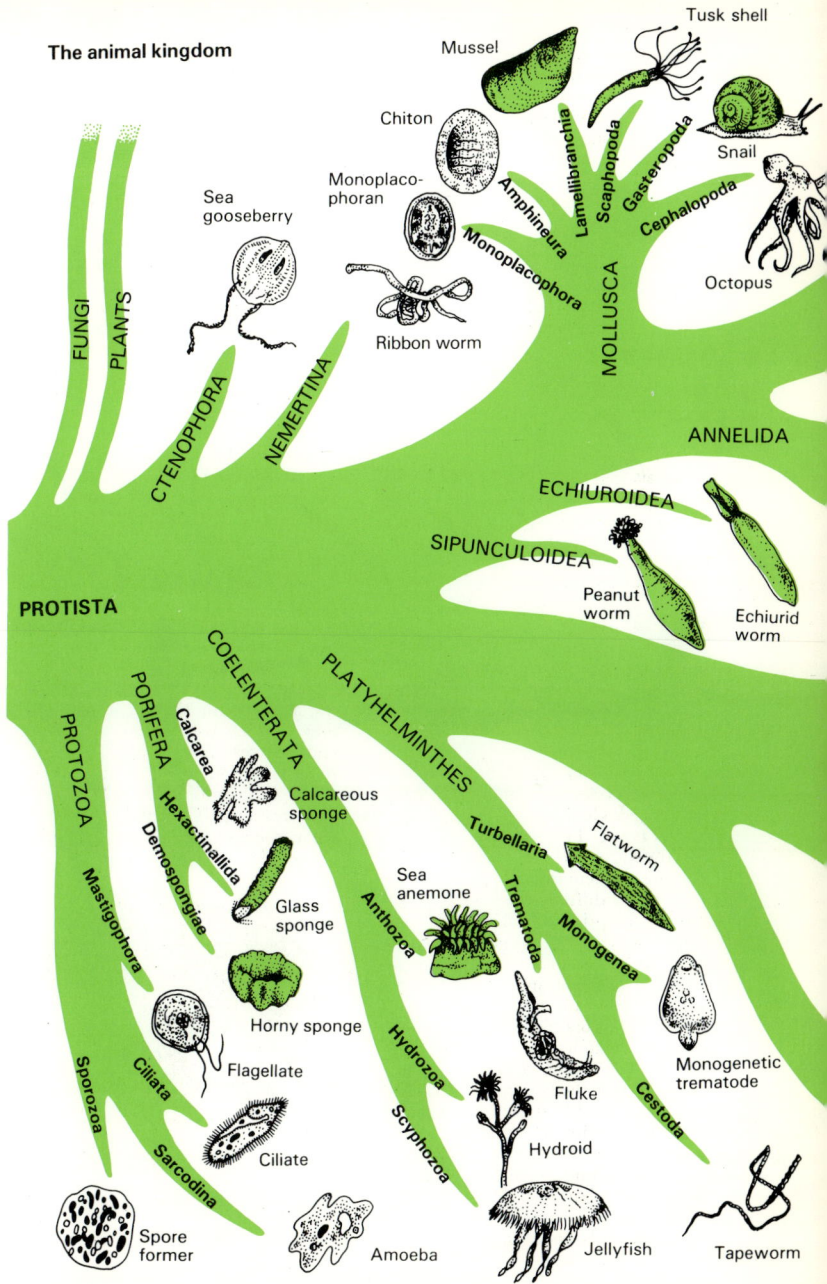

The animal kingdom

FUNGI

PLANTS

Sea gooseberry

CTENOPHORA

NEMERTINA

Chiton

Monoplaco-phoran

Monoplacophora

Ribbon worm

Mussel

Amphineura

Lamellibranchia

Scaphopoda

Tusk shell

Gasteropoda

Snail

Cephalopoda

MOLLUSCA

Octopus

ANNELIDA

ECHIUROIDEA

SIPUNCULOIDEA

Peanut worm

Echiurid worm

PROTISTA

COELENTERATA

PLATYHELMINTHES

PORIFERA

Calcarea

Calcareous sponge

Hexactinellida

Demospongiae

Glass sponge

Turbellaria

Flatworm

Sea anemone

Anthozoa

Trematoda

Monogenea

PROTOZOA

Mastigophora

Horny sponge

Hydrozoa

Fluke

Monogenetic trematode

Sporozoa

Ciliata

Flagellate

Scyphozoa

Cestoda

Sarcodina

Ciliate

Hydroid

Spore former

Amoeba

Jellyfish

Tapeworm

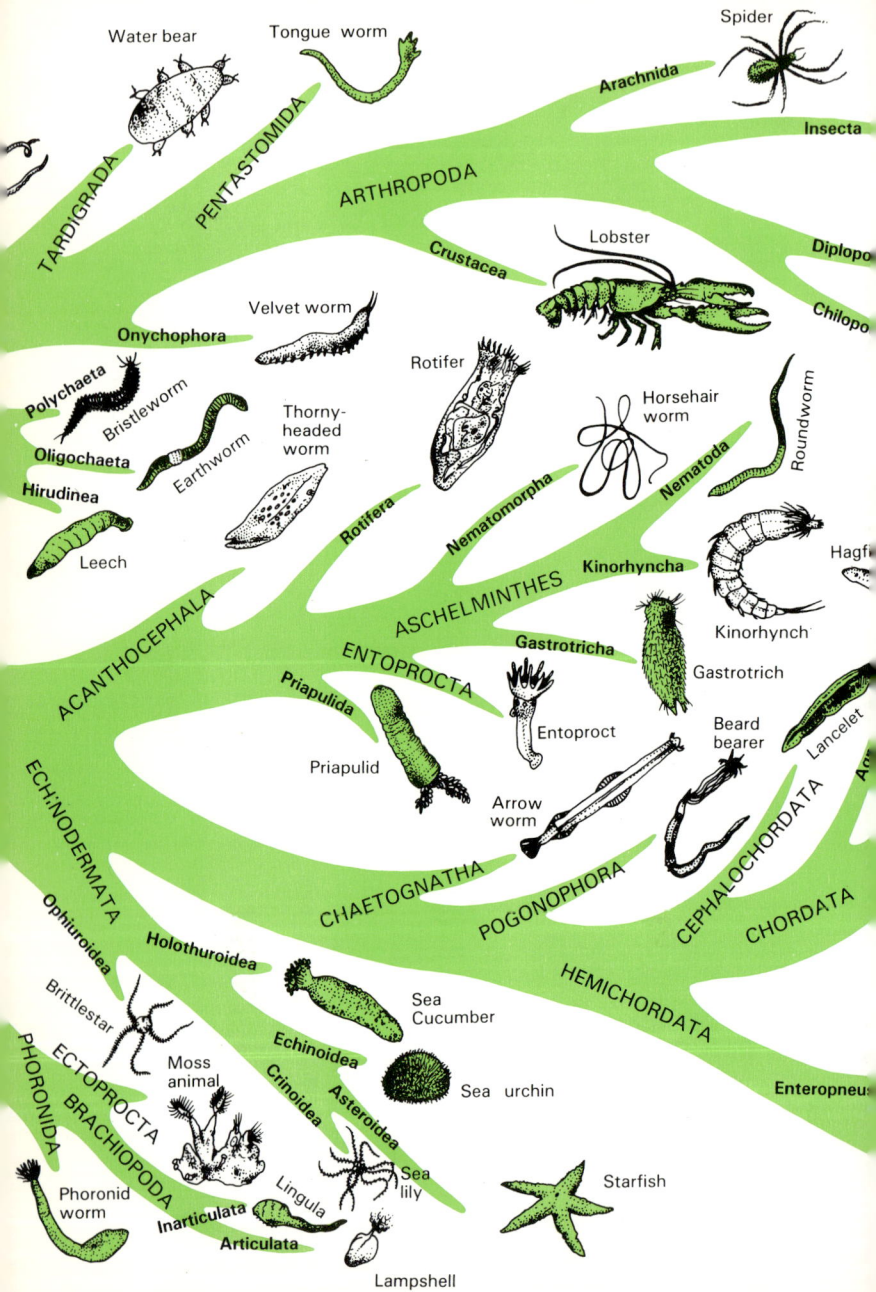

Water bear

Tongue worm

Spider

Arachnida

Insecta

TARDIGRADA

PENTASTOMIDA

ARTHROPODA

Crustacea

Lobster

Diplopo

Chilopo

Velvet worm

Onychophora

Rotifer

Horsehair worm

Roundworm

Polychaeta

Bristleworm

Thorny-headed worm

Oligochaeta

Earthworm

Rotifera

Nematomorpha

Nematoda

Hirudinea

Kinorhyncha

Hagf

Leech

ACANTHOCEPHALA

ASCHELMINTHES

Gastrotricha

Kinorhynch

ENTOPROCTA

Gastrotrich

Priapulida

Priapulid

Entoproct

Beard bearer

Lancelet

ECHINODERMATA

Arrow worm

CHAETOGNATHA

POGONOPHORA

CEPHALOCHORDATA

CHORDATA

Ophiuroidea

Holothuroidea

HEMICHORDATA

Brittlestar

Sea Cucumber

ECTOPROCTA

Moss animal

Echinoidea

Sea urchin

PHORONIDA

BRACHIOPODA

Crinoidea

Asteroidea

Enteropneus

Phoronid worm

Lingula

Sea lily

Starfish

Inarticulata

Articulata

Lampshell

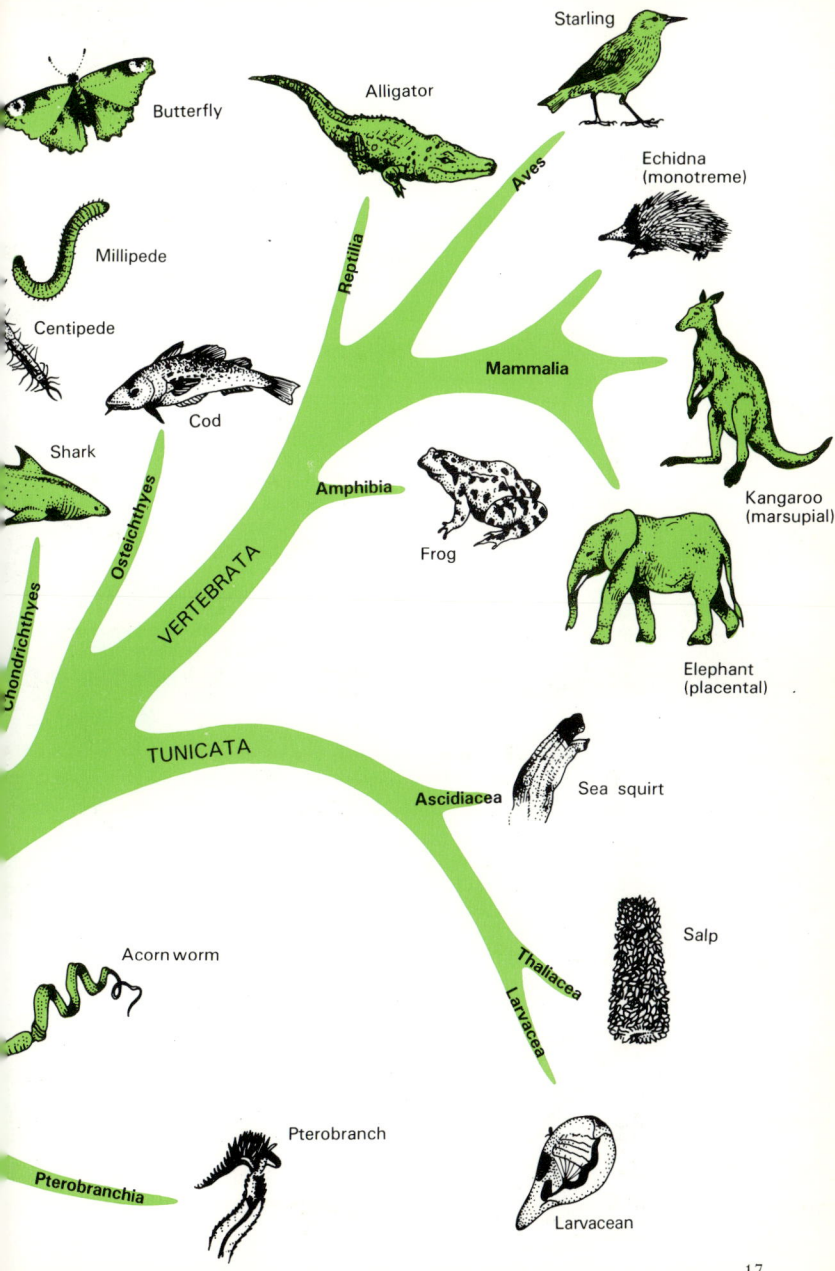

Butterfly

Alligator

Starling

Aves

Echidna
(monotreme)

Millipede

Reptilia

Mammalia

Centipede

Kangaroo
(marsupial)

Cod

Osteichthyes

Shark

Amphibia

VERTEBRATA

Frog

Chondrichthyes

Elephant
(placental)

TUNICATA

Ascidiacea

Sea squirt

Salp

Acorn worm

Thaliacea

Larvacea

Pterobranch

Pterobranchia

Larvacean

17

Giant anteater

Both have prehensile tails, with which they can grip the branches. All of the anteaters have long, slender tongues which they use to lick up ants and termites. They have no teeth.

Antelope A group of hoofed mammals belonging to the same family as the cows, sheep and goats. They have horns which are unbranched and which do not fall off each year. Both males and females may have horns. Antelopes live in Africa and Asia, often forming huge herds. GAZELLES, ORYXES, and ELAND are all antelopes.

Antenna (plural antennae). A sensory organ on the head of various arthropods . It can detect scents and vibrations and sometimes heat.

Anther The swollen tip of a stamen which produces the pollen.

Antheridium The body containing male sex cells in lower plants such as MOSSES and the prothalli of FERNS.

Antibody A protein which is produced in the blood of an animal in response to certain foreign substances, such as bacteria, that enter the body. The antibodies formed combine with the foreign body or antigen to render it harmless or more easily dealt with by the infected animal.

Antigen A foreign substance, e.g. a bacterium or virus, which invades the body of an animal and induces the production of antibodies.

Ant lion An insect resembling a dragonfly but more closely related to the lacewing fly. The name 'ant lion' refers to the young insects. They live where the soil is sandy and some species make little pits for themselves. They lie more or less buried at the bot-

tom of the pit and wait for an ant or some other insect to tumble down the loose sandy sides. They then pounce on it and eat it. Other species merely lie buried in the sand with just their great jaws exposed, ready to grab any passing insect. Ant lions live in most of the warmer parts of the world.

Aorta A major artery of man and other vertebrates which carries blood from the heart to the body.

Apetalous (of a flower) Without petals.

Aphid 'Greenfly' or 'blackfly' are two of the commonest of the many types of aphid. Also called plant lice, aphids are sap-sucking bugs and they do an immense amount of damage to crops by sucking sap and by spreading germs that cause plant disease. Most aphid species have both winged and wingless forms.

Apocarpous (of a flower) Having a number of separate carpels.

Apollo butterfly A group of mountain-dwelling butterflies related to the swallowtails. Found in Europe, Asia and North America, they are mostly white with black or red markings.

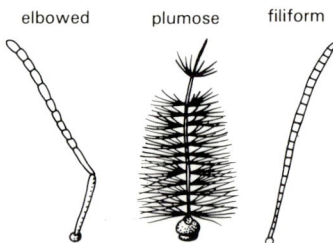

elbowed   plumose   filiform

Various types of antennae

Appendix Blind sac in the alimentary canal of mammals. It is a vestigial organ and of little importance to most animals.

Apple A member of the rose family, the apple is the best-known and most widely-grown fruit of temperate regions. It has been cultivated since prehistoric times. Apple trees are long-lived and easy to grow. More than 7000 varieties are cultivated.

Apricot A tree producing a golden fruit with a stone, native to China. Apricots are grown in Europe and North America. They belong to the plum family.

Apus A small freshwater crustacean which resembles the long-extinct trilobites in appearance and habits.

Arabian camel, *see* CAMEL

Arabian oryx, *see* ORYX

Archerfish

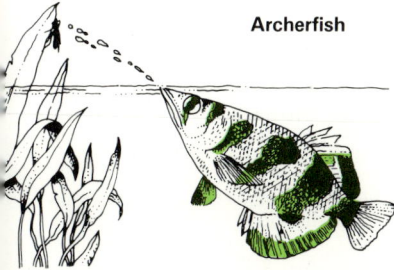

Arachnid A member of the arthropod class Arachnida which includes spiders, scorpions and mites. Unlike all other arthropods the arachnids have a pair of clawed pincers as the first appendage on the head. The next pair are either larger sensory palps or pincers and the remaining four pairs are walking legs. Arachnids have no antennae.

Arapaima A pike-like freshwater fish of South America. It is one of the largest freshwater fishes in the world, reaching a length of 2 metres or more and weighing over 100 kg.

Arborvitae A coniferous tree of Asia and North America. It produces sweet-smelling, durable timber.

Archegonium The flask-shaped structure that contains the female sex-cells of mosses, ferns and related plants.

Archerfish A small fish of muddy coastal waters and swamps in India and South-East Asia. Archerfishes have a remarkable way of obtaining insect food: they squirt a stream of water droplets at insects resting on overhanging plants and knock them into the water. Archerfishes have been known to hit insects from a distance of 2 metres.

Arctic fox A small fox with short ears and muzzle, giving it a somewhat cat-like appearance. It lives in the far north and its greyish yellow summer coat turns white in winter. Like the polar bear, the arctic fox has hairy soles on its feet. This helps it to get a grip on the snow and ice.

Arctic hare A large hare found on the treeless tundras of North America and in Greenland. In the northernmost part of its range, the arctic hare is white all year round; further south it has a grey back in summer; only the tips of the ears are always black.

Argali The largest of the wild sheep, the argali stands well over one metre at the shoulder and has spirally twisted horns up to 2 metres long. It is found in central Asia.

Argonaut A marine mollusc related to the cuttlefish and living in warm waters. The female, up to 30 cm long, lays her eggs in a wafer-thin shell. The male, which rarely reaches 1 cm in length, was identified by zoologists little more than 100 years ago.

Armadillo A group of armoured omnivorous animals ranging from Patagonia to the south-western USA. Largest is the giant armadillo of South America which reaches one metre in length. The three-banded armadillo is the only one able to roll up into a protective

Armadillo

ball. Other species include the nine-banded armadillo of Mexico and the fairy armadillo of the Argentinian plains.

Armoured catfish Freshwater South American fishes with whisker-like barbels and bony plates protecting part or all of the body. Armoured catfishes can live out of water for a considerable time. They can even travel overland from one river to another, dragging themselves along with their strong pectoral fins.

Army ant Also called driver ants or legionary ants, these insects are confined to the warmer parts of the world and have no permanent nests. The colonies move about in long columns, with large-jawed workers on the outside to defend them. Any animal which is unable to move out of the way of the column is eaten, no matter what its size. Males have wings and look very different from the other members of the colony. Their light brown bodies may exceed 2.5 cm in length and they are known as sausage flies.

Arrowhead An aquatic plant growing in quiet muddy waters. It has white flowers, arrow-shaped leaves and edible tubers.

Arrow poison frog One of the most powerful poisons in nature is contained in the skin of these South and Central American frogs. It has long been used by Indians for making poison-tipped arrows. Most species of arrow poison frog have bright colours which act as warning signals to would-be predators.

Arrowroot A tropical West Indian herb. Starch from its tuberous rhizomes makes an easily-digested food.

Arrow worm A little animal, up to 10 cm long, that floats in the plankton of the sea. The body is transparent except for a pair of tiny black eyes, and it is very difficult to see this creature unless it has just had a meal. Arrow worms have spiny jaws and they catch small animals by darting after them and spearing them.

Arteriole A small artery.

Artery A vessel leading blood away from the heart and into the tissues. For example: the pulmonary artery to the lungs, the renal artery to the kidneys and the carotid artery to the head.

Arthropod A member of the largest animal phylum, the Arthropoda. All arthropods have jointed limbs and are normally covered by a stout cuticle. The latter usually contains a horny protein called CHITIN. Like the annelid worms, from which they are probably descended, the arthropods show definite segmentation of the body. They include INSECTS, CRUSTACEANS, ARACHNIDS, BRISTLETAILS, CENTIPEDES and MILLIPEDES.

Artichoke, Globe A grey-green, thistle-like plant up to 90 cm tall. The fleshy scales of the flower-heads are eaten.

Artichoke, Jerusalem A tall coarse, sunflower-like plant up to 3.5 metres tall, native to North America. Its knobbly tubers are eaten.

Arum The name given to a family of mainly tropical plants in which the poker-like flower-spike or spadix is surrounded by a large sheath or spathe. Many give off unpleasant smells which attract pollinating flies. A number have sword-shaped leaves. The common arum or cuckoo-pint is a wild variety with poisonous, red berries.

Common arum (cuckoo-pint)

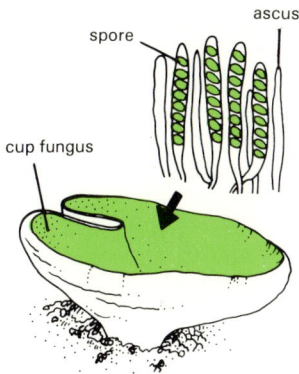

Ascomycete

Ascomycete The ascomycetes form one of the major groups of fungi. The fruiting bodies form spores inside cells called asci which explode and shoot the spores out when ripe. The CUP FUNGI are the most familiar ascomycetes.

Ascorbic acid Better known as vitamin C. It prevents scurvy and assists wounds in healing. It may be manufactured by some animals but is usually obtained directly from fruits and fresh vegetables.

Ascus The spore-producing cell of certain (Ascomycete) fungi. An ascus contains eight ascospores.

Asexual reproduction is the increase of numbers by simple division or budding without any form of mating or the fusion of two sex cells.

Ash A group of large trees found in Europe, Asia and North America. The common ash of Europe and south-western Asia is a majestic tree reaching a height of 40 metres. It has large leaves made up of 9–15 leaflets. It produces tough, elastic timber.

Asp A poisonous snake closely related to the adder and distributed over the warmer parts of Europe. Its colour ranges from greyish-brown to orange or red, and it rarely exceeds 60 cm in length. It can be distinguished from the adder by its turned-up snout and by its yellow eyes. The adder's eyes are coppery-red.

Asparagus An edible plant of Europe and a member of the lily family. Asparagus can take up to five years to produce its first crop. Only the young, fleshy stalks are eaten. Asparagus fern is an ornamental variety from southern Africa.

Aspen A species of POPLAR. Its leaves move in the slightest breeze.

Aspidistra A plant with long glossy leaves, native to China and Japan. It was an extremely popular indoor plant in dark Victorian houses because it tolerates poor light.

Ass The ass is a small kind of horse. Two species are known — one in North Africa and one in Central Asia. A race of the African species has been domesticated for a long time and we know it as the donkey. There may well be no truly wild asses left in Africa today, but the Asian species survives in several countries.

Assassin bug There are more than 3000 kinds of assassin bug, so called because of the violent way they seize and kill their insect prey. Assassin bugs have a powerful curved beak which is thrust into the prey and which then sucks up their juices. Many assassin bugs mimic their prey and mix with them freely without causing alarm. Most species live in the tropics.

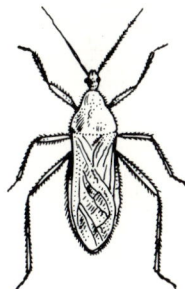

Assassin bug

Assassin fly, *see* ROBBER-FLY

Assimilation The process of converting simple food materials into complex tissues of an organism.

Association The method by which animals 'remember'. An encounter leading to an unpleasant experience is avoided at a future date whereas one leading to a reward will be repeated. In this manner animals can be trained.

Atlas moth One of the largest moths in the world, with a wing span up to 25 cm, the atlas moth is found in India and other parts of South-East Asia. It is mainly brown, but there is a transparent spot on each wing. The atlas moth belongs to the same family (Saturniidae) as the European emperor moth and the Indian tussore silk moths, but it does not provide useful silk.

Atria, see HEART

Aubergine (or egg-plant) A large plant producing purple egg-shaped fruits. It is a member of the potato family and a native of southern Asia, but the aubergine is grown as a vegetable in other warm places, and in greenhouses.

Auditory Concerning hearing.

Auk A group of black and white, duck-like seabirds with short, narrow wings. Most are strong fliers and spend most of the time at sea, only coming ashore to breed. Auks include the little auk, PUFFIN, RAZORBILL and GUILLEMOT.

Auricle, see HEART

Autolysis The self-destruction of cells in an organism, caused by the activity of the cell's own enzymes.

Autonomic nervous system Part of the nervous system responsible for gland secretion and stimulation of muscles not directly controlled by the will (e.g. musculature of the heart and digestive systems).

Autotomy Deliberate shedding of part of a limb or tail when seized by an enemy. Limbs shed in this way usually grow again.

Autotrophic organisms are able to manufacture their own organic food from inorganic materials using energy from an outside source. Most green plants are completely autotrophic, making organic foods from carbon dioxide, water, and mineral salts, using the energy of sunlight which is in some way fixed by the chlorophyll. Some bacteria are also autotrophic, using energy obtained by the oxidation of inorganic substances. The nitrifying bacteria of the soil, and the iron bacteria are examples.

Auxin A plant growth substance or hormone.

Avadavat A small strikingly-coloured bird of South-East Asia. The male has red and black plumage dotted with white spots.

Avocado (or alligator pear) A tropical tree bearing large, pear-shaped fruits. The fruits are eaten in savoury dishes.

Avocet A graceful black and white wading bird with a long upturned beak. Avocets are found along shores and by rivers in many parts of the world. The upturned beak allows the avocet to feed on floating animals and seeds. The beak is held just under the surface, so that the end is more or less horizontal. The bird then walks along, sweeping its head from side to side and scooping material from the water.

Axil The angle between a leaf or bract and the stem on which it grows.

Axis deer The two species of axis deer live in India and South-East Asia. One species, known as the chital or spotted deer, is a small, dainty animal with a white-spotted red coat. The other axis deer is quite different. Its squat pig-like appearance has given it the name hog deer. Both species of axis deer have slender antlers, up to one metre long in the chital, but only 30–40 cm in the hog deer.

Axolotl A newt-like creature about 15 cm long which breathes by means of feathery gills just like a young newt or salamander. But, whereas the young newt eventually develops lungs and leaves the water, the axolotl normally retains its juvenile appearance throughout its life. Only under unusual conditions does it turn into an adult salamander. Axolotls live in Mexico.

Axon The part of a nerve cell along which an impulse or signal is carried to a neighbouring cell.

Aye-aye A squirrel-like animal, closely related to the lemurs, found only in the forests of Malagasy (Madagascar). Despite protection the aye-aye is in danger of extinction.

Azalea A deciduous flowering shrub native to China and North America, related to rhododendrons. Azaleas have fleshy green leaves and colourful blooms.

B

Babbler The name given to various unrelated birds with a loud and varied call. Most live in south-eastern Asia but there are babblers in Australia and one species, the wren-tit, lives in North America. Babblers live near the ground in the forests, searching for berries and insects in the undergrowth.

Baboon A mainly ground-living primate of Africa which lives in highly organized family groups called troops. Baboons forage on a wide variety of food during the day and sleep in the safety of trees at night. Lions are their chief enemy.

cocci

spirilli

Bacteria

bacilli

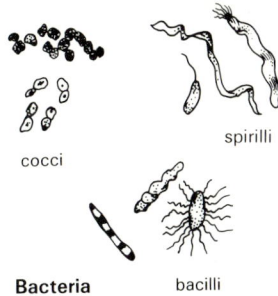

Baboon

Backswimmer A group of water bugs which swim upside down, using their long hindlegs as oars. Under water, they breathe from an air bubble trapped against the abdomen by fine hairs. Backswimmers are fierce predators of small fishes and can inflict a painful bite if handled.

Bacteria Minute, single-celled organisms, so small that they cannot be seen with the naked eye. The largest are only about 0.01 mm long. Many are essential to the processes of decay, helping to break down the dead remains of plants and animals and releasing materials that plants can absorb through their roots. Others have the ability to use atmospheric nitrogen in the building up of nitrates — mineral salts that are vital for healthy plant growth. Many of the bacteria present in the gut of animals break down food materials and thus provide substances that the animal would otherwise be unable to obtain. The most obvious effect of bacterial activity, however, is disease. Only a small proportion of the species of bacteria are harmful, but even so the list of diseases that they cause is a lengthy one. Bacteria may be arranged into three groups according to their shape: rod-shaped forms (bacilli), spherical forms (cocci), and curved or spiral forms (spirilli).

Bacteriology The study of bacteria.

Bacteriophage A virus that can kill bacteria.

Bactrian camel, *see* CAMEL

Badger The European badger is a short stocky animal about one metre long. Its coat looks grey from a distance, but the hairs are actually black and white. The most striking feature is the black and white striped head. Badgers use their strong front claws to dig extensive underground homes, called sets, usually in woodlands. At night they follow regular tracks in search of food. Badgers are carnivorous creatures, but they also eat a variety of plant food. The American badger is quite similar, but lacks the black and white striped head of its European cousin.

Bagworm The name given to various moths whose caterpillars make little cases from silk and plant fragments. Most of them feed on trees, and some are considerable pests. When ready to pupate, the bagworm larva fixes its

Bald eagle

case to a twig and pupates inside it. The female adult is wingless and usually remains inside the case. The male is fully winged and active.

Bald eagle This striking bird used to be found all over North America and it was adopted as the emblem of the United States. Today it is extremely rare. It is not really bald; it gets its name from the pure white feathers on its head and neck, which contrast vividly with the brown feathers on the rest of the body. The bald eagle feeds mainly on fish.

Baleen The name given to the horny plates of keratin that hang from the upper jaws of the toothless whales. The plates filter out the small crustaceans on which the whales feed.

Balsa A tall tree producing the lightest commercial timber. It grows in western America from Mexico to Bolivia. Some of the wood weighs less than cork.

Balsam Also known as touch-me-not, this is a fleshy-stemmed flowering plant whose ripe seed-pods burst at the slightest touch. Tropical species of balsams are grown in hothouses.

Bamboo A group of giant grasses growing in warm climates. Bamboos have tough, tubular stems, growing up to 37 metres tall, and long blade-like leaves. Bamboo is used as timber and food, and for making cloth and rope.

Baobab

water enters through mouth

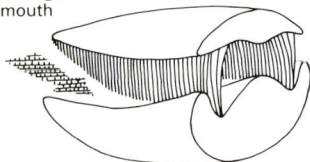

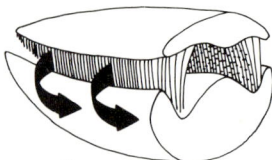

water filters out through baleen plates

Baleen

Banana A tropical plant producing large clusters of long yellow fruit. Its leaf-stalks grow together to resemble a woody trunk. There are many varieties, whose fruits vary in sweetness. Some particularly starchy forms are called plantains. Cultivated varieties generally have no seeds and are grown from cuttings.

Banded anteater, *see* NUMBAT

Bandicoot A group of rat-like marsupials found in Australia and New Guinea. They are mainly nocturnal animals, most spending the day in simple grass nests, but rabbit bandicoots dig burrows.

Banyan An Indian fig tree whose branches send out roots which descend to the ground and form trunks. A large banyan can look like a small forest because it has so many trunks. The banyan often starts life on the branch of another tree, from a seed dropped there by a bird.

Baobab An African tree with a very thick trunk, up to 15 metres in diameter. It has gourd-like fruit, called monkey bread.

Barbary ape This is not an ape at all, but a monkey of the macaque family. It is about 60 cm long and has an extremely short tail. Barbary apes live in North Africa and a small colony lives in Gibraltar.

Barbary sheep Found in the mountains of north-western Africa, this sheep has a long mane, particularly in the male: it grows on the underside of the neck and reaches down the front legs.

Barber fish These fishes come from a wide range of families and are also called cleaner fishes. They feed on the parasites attached to other fishes. While this takes place, the fish being cleaned co-operates with the barber, even allowing it to enter the mouth and gill cavities.

Barberry A low spiny shrub producing yellow flowers and acid red berries. Barberries grow in temperate northern countries, including North Africa.

Barbet A group of colourful birds related to woodpeckers. They live in the tropical forests of America, Africa and Asia, feeding mainly on fruit.

Bark A protective tissue of dead cells that covers the stems and branches of woody plants. In some species, oak and elm for instance, it appears rough and fissured. In others, such as beech, it is smooth and thin, while in yet others it is scaly (larch and sycamore). But the bark seen on the outside of a tree is not the whole of the structure: there is an inner region of bark which is much paler and which contains living cells. The cells of this inner bark make up the phloem — the tissue that conducts food through the plant.

Bark beetle These insects include a variety of small species whose larvae live by tunnelling just under the bark of trees. The tunnelling activity may separate the bark completely from the wood and, if larvae are very numerous, they can kill the trees. But many species merely make small tunnel systems, doing little real harm. The elm-bark beetle carries Dutch elm disease.

Barley An important CEREAL grain grown in all temperate countries. It is used as human and animal food, and to make malt. It is shallow-rooted, hardy and grows quickly. Barley is one of the oldest plants cultivated by man, being grown probably 8000 years ago in the Nile valley. Barley resembles wheat in appearance, its seeds growing in long clusters along a central spike.

Barnacle Small seashore creatures whose white tent-like shells coat rocks and pier legs. When the tide is out, the plates of the shell close up; when the tide returns the plates open and the animal starts to feed. Feathery arms come out and rhythmically comb food particles from the water. Despite their unusual appearance, the barnacles are crustaceans and are therefore related to crabs and shrimps. The young barnacles swim freely before settling down and making a shell.

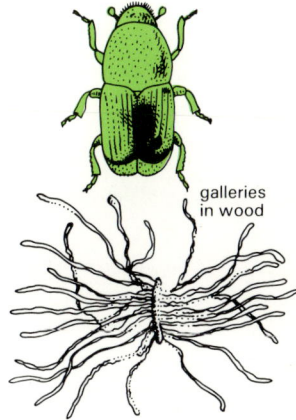

galleries in wood

Bark beetle

Barnacle goose A black and white goose with a conspicuous white face. Flocks appear in Europe in autumn, and they disappear again in the spring. Not until this century were their summer breeding grounds discovered in the far north — in Greenland and Spitzbergen.

Barn owl The most widely distributed of all land birds, the barn owl is found in all continents except Antarctica. It is very pale in colour and has probably

Barn owl

been responsible for many ghost stories. It has a white round face. Barn owls live in hollow trees and old buildings, such as barns, and they eat mainly rats and mice.

Barracuda A group of pike-like fishes with vicious-looking teeth. The larger species may reach 2.5 metres in length. Barracudas feed mainly on herrings and other surface-feeding fishes. They are found in the Atlantic and Pacific Oceans and the Black Sea.

Bar-tailed godwit, *see* GODWIT

Basidiomycete The basidiomycetes form one of the major groups of fungi. The fruiting bodies form spores on the

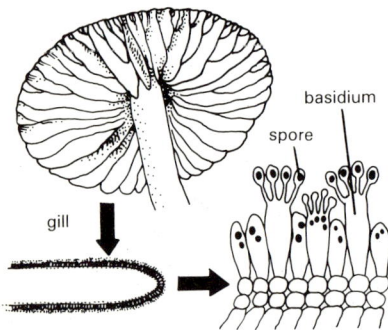

Basidiomycete

outside of club-shaped cells called basidia. All the umbrella-shaped toadstools are basidiomycetes, as are puffballs, bracket fungi, parasitic rusts and many others.

Basidium (plural basidia) The spore-producing cell of mushrooms and related fungi (Basidiomycetes). It carries four basidiospores on its surface.

Basil An aromatic herb used for flavouring food. Sweet basil, the most popular variety, originated in India.

Basilisk A group of lizards living in tropical America. They are up to 60 cm long, and the males have a crest on the head and back. The animals always live near water and they can actually skim across the surface using only their back legs. They will also run on land in this semi-erect position. Basilisks eat small animals and insects.

Basket star A marine animal related to starfishes. The central disc has five major arms, but these fork many times. Basket stars live on the seabed and capture small creatures with their branching arms.

Basking shark One of the largest fishes, the basking shark may reach 12 metres or more in length. It is most common in the North Atlantic. Despite its great size, the basking shark has tiny teeth and rarely attacks man. It feeds on plankton, which it strains from the water with its very large gills.

Bass There are many kinds of fish called bass, living in both salt and fresh waters. Some of the larger marine species may weigh well over 50 kg. The true bass lives in the Mediterranean and the eastern Atlantic. The stone bass, or wreckfish, likes to live in and around ship-wrecks. It is an aggressive fish, reaching a length of 2 metres.

Basswood (or American lime) A North American tree growing up to 37 metres tall. It produces light soft timber. Basswood has large heart-shaped leaves and yellow flowers.

Bat A flying mammal of the order Chiroptera. The 'wings' of a bat are elaborate folds of skin supported by its elongated fingers and its arms and legs. Bats are nocturnal creatures with extremely poor vision. They navigate at night using a kind of radar system, emitting high-pitched sounds which bounce back from nearby objects and are picked up by the bats' large ears. This radar is so sensitive that the bats can use it to detect and catch the flying insects on which most of them feed. Some large bats, known as FRUIT BATS, feed mainly on fruit. Apart from their modifications for life in the air, the bats closely resemble the insectivores.

Bat-eared fox A small fox of southern and eastern Africa with a jackal-like face and long ears measuring up to 11 cm.

Bath sponge The horny skeletons of several species of SPONGE are familiar in their role of the traditional bathroom sponge. These simple animals are found in warm seas, particularly the Mediterranean, down to 200 metres. Today nearly all bath 'sponges' are made of synthetic materials.

Bay tree The name of several different trees. The bay of southern Europe is an evergreen of the laurel family. It has aromatic leaves used for flavouring, and by the ancient Greeks to crown heroes. Sweet bay is an American species of magnolia.

Beaked whale A family of toothed whales with a distinct dolphin-like beak. The most common and largest are the bottle-nosed whales.

Bean A group of plants of the pea family, producing large seeds in pods. The seeds and sometimes the unripe pods are used as food. The plants

Pipistrelle bat

enrich the soil with nitrates, produced by bacteria in their roots. Varieties include French, broad, runner, lima, mung, kidney and soy beans.

Bear A large mammal belonging to the family Ursidae. Bears are plantigrade animals, i.e. they walk (like man) on the soles of their feet. They have powerful limbs with strong claws, small round ears and a short tail almost completely hidden in the fur. Bears are found in most parts of the world with the exception of Africa and Australia. They include the BLACK BEAR, BROWN BEAR, POLAR BEAR, SLOTH BEAR and SUN BEAR.

Bear-cat, *see* BINTURONG

Beaver The second largest rodent in the world, the beaver is up to 1.4 metres long, including the tail. It weighs up to 35 kg. There are two species, very much alike. One lives in northern Europe and the other in North America. Beavers always live

Beaver

beaver lodge

near water and they make their homes in river banks or in 'lodges' built of branches in the middle of a pond. Even the pond is made by beavers. They use their chisel-like front teeth to cut down small trees which they then use to build a dam across a stream. The beavers feed on bark, which they strip from trees. The front feet are clawed, for digging, and the hind feet are webbed. Beavers are excellent swimmers and they use their broad tails as rudders.

Bed-bug One of the many insect pests that man has acquired during his history. It is a small flat brownish insect that feeds by sucking blood. It has no wings and it hides in crevices or among bedclothes during the daytime.

Bedstraw Any of a group of herbs with square stems and whorls of leaves at each stem joint. The stalks were once used as straw for mattresses.

Honey bee

Bee A group of normally furry insects belonging to the order Hymenoptera. Many bees are social insects. The most highly organized of all is the HONEY-BEE whose colonies last year after year. The BUMBLE-BEE has a smaller colony which lasts for one year only. In the solitary bees the young leave the nest when they are mature. Bees all feed on pollen and nectar and play an important role in the pollination of flowers.

Beech Any of ten species of large deciduous trees, native to North America, eastern Asia, and Europe. Beeches

have dense foliage, and few plants grow under them. The timber is tough and much used for furniture.

Bee-eater The bee-eaters are very spectacular birds distributed over most of the warmer parts of the Old World. They are brightly coloured and they have curved beaks. They nest in sandy banks and their nesting colonies may contain thousands of birds. They feed on insects, including bees and wasps, which they catch in flight. They also destroy large numbers of locusts.

Beet A plant with large roots used as food. Red beetroots are cooked as a vegetable; white sugar-beets yield sugar; mangel-wurzels or mangolds are grown as cattle-feed.

Beetle With more than 250,000 known species, beetles form the largest of the insect orders — the Coleoptera. They range in size from less than 0.5 mm in length to the heaviest of all insects, the African Goliath beetle weighing about 100 grams. Beetles normally have two pairs of wings, the front pair being modified to form hard cases (elytra) which protect the rear wings used for flying. Most beetles can fly well, but their time is largely spent on the ground or on vegetation. The order contains plant-feeding, carnivorous, scavenging and parasitic forms.

Begonia Any of 900 species of plants grown for their showy flowers. The waxy flowers range from white to red. The leaves are sometimes tinted pink or red and the stems are rather succulent.

Belladonna lily, *see* AMARYLLIS

Bell animalcule A group of protozoans with a cup-shaped body attached to a stem. Some are anchored to underwater objects. Others join together in large, jelly-like colonies.

Bell bird The name given to several unrelated species of bird living in the southern hemisphere whose voices have a metallic ring.

Bellflower (or campanula) Any of 250 species of plants with bright, bell-shaped flowers. The blooms often hang down from the stems and are generally blue. Canterbury bell is a garden variety.

Beluga The belugas or white whales are related to the dolphins and por-

Bighorn

poises. The young animals are grey, but the skin gradually becomes white by reduction of the black pigment. Belugas usually grow to about 4 metres in length. They live in and around the Arctic Ocean.

Benthos Term used to cover organisms living on the sea floor as opposed to those swimming and floating in the water. It includes sea anemones, starfishes, molluscs and many others.

Berry A fleshy fruit usually containing many seeds. Examples include tomato, gooseberry and orange.

Betel nut The fruit of the Asian betel palm. Many people of southern Asia chew the betel nut as a narcotic stimulant.

Betony A small plant of Europe, North Africa and Siberia used by herbalists as a cure for fevers. It has heart-shaped leaves and deep red or purple flowers.

Bewick's swan, *see* SWAN

Bichir A primitive freshwater fish of tropical Africa whose body is covered with an armour of scales. The bichir has an air bladder connected with the throat, and it is able to breathe air in addition to obtaining oxygen through its gills. The front fins can be used as props, almost like legs, and the bichir may well represent one of the stages through which land animals passed during their evolution from fishes.

Biennial A plant that completes its life cycle in two seasons. Examples include the carrot and beetroot. The first season is involved in the production and storage of food, and in the second year this food store is used up in the production of flowers and seeds. The plant then dies.

Bigheaded turtle A heavily armoured freshwater turtle of south-eastern Asia whose head is very large in relation to its body and cannot be withdrawn into the shell.

Bighorn A wild sheep living in the Rocky Mountains. The males have enormous horns which curve around the back of the head and come forward again to the eyes. The horns are marked off into segments, each of which represents one year of the animal's life.

Big-scale, *see* TARPON

Big-tree, *see* SEQUOIA

Bilberry A small spreading shrub of Europe and northern Asia. It produces small, edible blue-black berries. The American blueberry is a relation.

Bile A secretion of the vertebrate liver which is passed to the intestine where it aids in the digestion of fats. Bile is usually formed in a small sac called the GALL-BLADDER. It also contains various waste products which are ejected with the faeces.

Binary fission

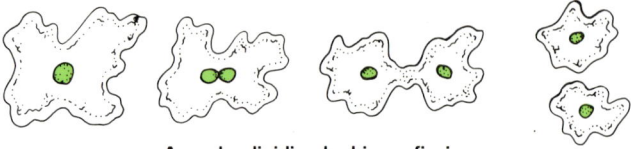

Amoeba dividing by binary fission

Binary fission Simple division of a cell into two.

Bindweed Any of a group of plants that twine their long thin stems around other plants. Hedge bindweed has big bell-shaped white flowers, black bindweed has small green flowers, and field bindweed small pink or white blooms. The leaves may be heart-shaped or arrow-shaped.

Binocular vision In many animals the eyes are in the side of the head. Each eye has a distinct field of view and vision is said to be monocular. But other animals (e.g. primates) have binocular vision: the eyes are in front of the head and their fields of view overlap. This enables the animal to judge distances more accurately.

Binturong The binturong or bear-cat is a tree-dwelling carnivore found in South-East Asia. It is about one metre long and has black fur. The binturong is clumsy on the ground, but it is an excellent climber. Its tail, which accounts for nearly half the animal's length, can be wrapped around

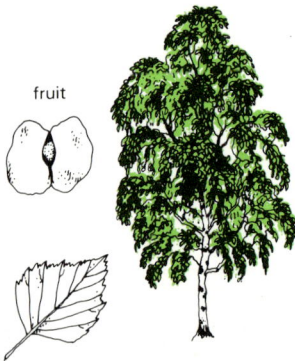

fruit

Silver birch

branches as a brake or a fifth limb. The binturong feeds on birds and small mammals, together with a fair amount of fruit.

Biochemistry Study of the chemical processes which take place in cells and tissues.

Biological control The use of one organism to control the population of another — usually a pest. One example concerns the prickly pear cactus. Introduced into Australia in the 19th century, this plant spread so rapidly that it became a serious pest. It was eventually controlled by the introduction of a moth whose caterpillar feeds on this cactus.

Biology The study of living things.

Bioluminescence The production of light by a living organism. Many animals, notably the GLOW-WORM and the FIRE-FLY, produce their own light but among plants only the lower ones — certain toadstools, moulds and bacteria — exhibit this phenomenon.

Birch Any of about 40 rather slender trees of the north temperate zone. They have thin, peeling bark, small leaves, and durable white wood. Their flowers are carried in slender catkins.

Bird A warm-blooded vertebrate of the class Aves that is unique in possessing feathers. Like the reptiles from which they are descended, birds lay shelled eggs. Another reptilian characteristic is the presence of scales on the legs and the feet. The forelimbs are modified as wings and the body of a bird is designed for flying: many of the bones of the skeleton are hollow and the large flight muscles may account for one-fifth of the entire body weight. Birds have no teeth but a tough horny beak. The food is not chewed but is stored in the crop after it is swallowed and is later ground up in the muscular gizzard which may contain stones.

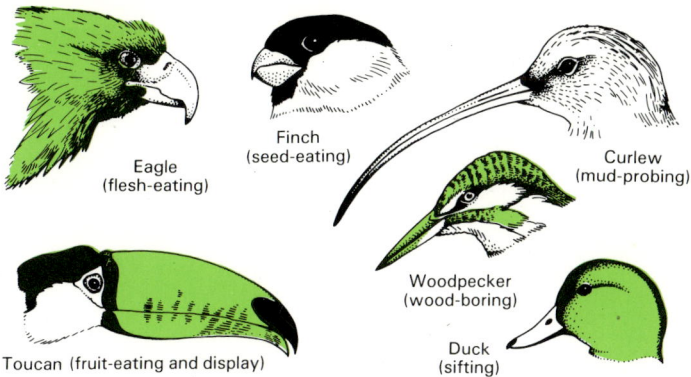

Various types of beaks and their uses

Eagle (flesh-eating)

Finch (seed-eating)

Curlew (mud-probing)

Woodpecker (wood-boring)

Toucan (fruit-eating and display)

Duck (sifting)

Bird-eating spider This large spider measures almost 20 cm across. It lives in the jungles of the Amazon basin. It does not spin a web, but catches birds and small mammals by dashing after them. The bird-eating spider has few enemies because it is covered with poisonous hairs.

Bird of paradise A group of birds found mainly in the forests of New Guinea and surrounding islands. The female birds are rather dull in colour, but the males of most species have beautiful plumes which they use in the most elaborate courtship displays.

Bird-of-paradise flower A South African plant with brilliant orange and purple flowers. It is related to the banana and has banana-shaped leaves.

Birdwing The birdwings are magnificent butterflies living in the forests of South-East Asia. Some have wingspans exceeding 25 cm. The females are a little larger than the males, but they are less colourful. The males have velvety black wings, marked with beautiful shining greens and blues.

Bishop bird A group of sparrow-like weaver birds found in many parts of Africa. The female has a dull brownish plumage throughout the year, but the male acquires brilliant plumage during the mating season. Some species are popular cage birds.

Bison A massive, ox-like creature weighing well over one tonne. The largest ones stand almost 2 metres at the shoulder. The hair on the head, neck, shoulders, and forelegs is long and shaggy. There are two species: the European bison, or wisent, and the American bison. The latter, often

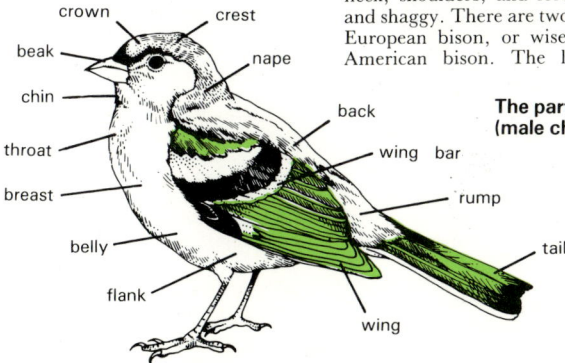

The parts of a bird (male chaffinch)

crown

crest

beak

nape

chin

throat

back

breast

wing bar

belly

rump

flank

tail

wing

wrongly called the buffalo, has longer hair than the wisent. Both species were almost exterminated during the last 100 years, but numbers have now increased again due to proper game management.

Bitterling A minnow-like fish, about 7 cm long, living in the streams of central and eastern Europe. The female lays her eggs in the breathing tube of a freshwater mussel where they are fertilized by the male and where they stay until they hatch.

Bittern A group of heron-like, marsh-dwelling birds found in many parts of the world. The common bittern is noted for the booming call of the male during the mating period and for the bird's habit of holding its head vertically when disturbed, the dark stripes on the throat making it virtually invisible against the background of reeds.

Bittersweet (or woody nightshade) A vine-like European plant. It climbs in hedges and has star-like purple flowers with yellow centres. American bittersweet has red and yellow flowers, and is not related.

Bivalve One of the large group of molluscs that have two parts, or valves, to their shell. The valves are hinged along one edge by a horny ligament and often by interlocking teeth as well. Examples include cockles, mussels, oysters and razor shells.

Black bear There are five species of black bear, living in the Americas and southern Asia. They are much smaller than the grizzly and the other brown bears, weighing only about 200 kg. The American black bear is very common in North America, especially in the national parks. Like its relatives, it is a good tree climber. The black bears eat a wide variety of food, including fruits, berries, rodents and insects.

Blackberry (or bramble) A thorny, trailing plant with white or pink flowers. The fruits, called blackberries, are actually clusters of small DRUPES. There are hundreds of kinds.

Blackbird The name given to a variety of unrelated birds in different parts of the world whose plumage is mainly black. The European blackbird is a member of the thrush family. It is a familiar garden bird with a beautiful

song and a clucking alarm note. The male is glossy black with a bright orange bill. The female is brownish with speckled underparts and can be mistaken for a thrush. The American blackbirds are insect-eating birds related to orioles.

Blackbuck Also known as the Indian antelope, the blackbuck is found over much of India and Pakistan. It is about one metre long and the male has remarkably twisted horns. It is one of the fastest land animals and has been credited with speeds of 80 kp/h. It can also leap well. The blackbuck is one of the few antelopes in which males are dark brown above, while the does and young animals are yellowish brown.

Blackfly, *see* APHID

Black Molly This little fish, a favourite with aquarists, is a black variety of the sailfin or related fishes which live in the rivers of Central America. The black varieties crop up from time to time in nature and they are bred artificially by aquarists. The fishes can be recognized by the large sail-like dorsal fin of the male.

Black panther, *see* LEOPARD

Blackthorn (or sloe) A shrubby tree with black bark and long thorns. Its fruits, called sloes, are like small, dark, and very bitter plums. The blackthorn's white flowers appear very early in spring — long before the leaves open. The leaves are oval-shaped and borne on reddish stalks.

Black widow This infamous spider has a number of sub-species distributed over the warmer parts of the world. The North American sub-species is noted for the female's powerful venom which produces severe pain and paralysis in a person. It can sometimes cause death, but fatal bites are rare. The female spider is about 1 cm long and shiny black in colour. There is a red 'hour-glass' mark on the underside of her spherical abdomen. The male is much smaller. The name of the spider comes from the belief that the female always eats her husband after mating, but this does not always happen.

The American black bear is highly popular in National Parks, where it begs food from the motorists.

Blue whale — showing size in comparison with an elephant

Bladder The urinary bladder where urine is stored is usually referred to as *the* bladder but, strictly, any thin walled sac is a bladder — e.g. GALL-BLADDER and SWIM-BLADDER.

Bladder-nut A deciduous shrubby European tree, with each leaf divided into five leaflets. Its seeds are in bladder-like capsules.

Bladderworts A group of carnivorous water plants with no roots. The leaves and stems are covered with tiny bladders, each of which acts as a trap for minute insect larvae and other aquatic animals.

Blenny A widespread group of small fishes common in shallow coastal waters. They are often found on rocky shores when the tide is out, sheltering under seaweed and darting across rock pools. Most blennies have large eyes which give them an intelligent expression.

Blesbok Small sturdy antelopes standing about 1.25 metres at the shoulder. There is a white blaze on the face and a white patch at the base of the tail. The short horns diverge from each other near the base and then become more or less parallel. Blesbok were once in danger of extinction. Many now live a semi-domesticated life on South African farms.

Blind snake This name applies to several hundred small, non-poisonous snakes living throughout the warmer parts of the world. They are worm-like creatures, usually only about 15 cm long, although some species reach one metre. They are not completely blind, but their eyes are very small and they can do no more than distinguish light from dark.

Blind spot The part of the eye where the optic nerve leaves the retina. Here there are no light-sensitive cells so that an image received there is not transmitted to the brain.

Blister beetle Blister beetles are most common in warm, dry climates. They get their name because their bodies contain a substance called cantharidin, which produces blisters on the skin of anyone touching them. The only British species has a shiny green body and is known as the Spanish fly.

Blood A fluid that circulates in the body of all higher animals, carrying food and oxygen to the tissues and removing waste products from them. It is circulated through a series of vessels and spaces by the pumping action of the heart. Blood consists of a colourless fluid called PLASMA which contains millions of tiny cells or BLOOD CORPUSCLES. The precise composition of

the blood varies among the different groups of animals.

Blood corpuscle The cells of the blood. There are two types found in vertebrate blood: red corpuscles and white corpuscles. Red corpuscles contain a pigment called HAEMOGLOBIN which carries oxygen through the body. White corpuscles are involved in the destruction of bacteria and any other parasites or foreign substances which invade the body. Certain of them are responsible for the production of ANTIBODIES.

Blow-fly The general name given to bluebottles and greenbottles. The flies are particularly unwelcome in kitchens since they lay their eggs on meat. The eggs hatch in about one day and the larvae or maggots feed on the decaying or 'fly-blown' meat.

Bluebell (or wild hyacinth) A bulbous plant of the lily family, a native of western Europe. It has blue (sometimes white or pink) flowers and long thin leaves. The Scottish bluebell is a HAREBELL.

Bluebottle, *see* BLOW-FLY

Bluefin tuna, *see* TUNA

Bluefish A voracious fish weighing up to 10 kg which lives in most of the warm seas of the world. It swims in large shoals, and attacks all kinds of fishes with astonishing ferocity. The bluefish occurs in exceptionally large numbers in the western Atlantic and is caught commercially for its tasty flesh.

Blue-green algae Small, often microscopic organisms in the kingdom PROTISTA. They form a scum on rocks, in soil, and on still water.

Blue hare, *see* MOUNTAIN HARE

Bluejay, *see* CUCKOO-SHRIKE

Blues A world-wide group of small butterflies, most of which are blue in colour, though the females are often partly or wholly brown. The caterpillars of blues often secrete a sugary fluid called honeydew which is drunk by ants.

Blue whale This is the largest animal that has ever lived. It can reach 30 metres in length and nearly 150 tonnes in weight — as much as 30 elephants. Yet this huge creature feeds on tiny planktonic animals which it strains from the water through plates of a horny material called baleen, or whalebone, in its mouth. During the summer, blue whales live in the polar seas, but they move to warmer seas in the winter, and their calves are born in these warmer regions. The calves are over 7 metres long when they are born. The species is in great danger of dying out because whalers have killed so many of them.

Boa Boas are large snakes found in the warmer parts of America, especially the Amazon basin. They kill their prey by coiling around it and suffocating it, but they are not poisonous. Large lizards, birds and some mammals are eaten. Best known of the boas is the boa constrictor, which sometimes reaches a length of about 4 metres. The ANACONDA is the largest of the boas. There are also some much smaller boas, known as sand boas, in Africa and Asia.

Boar, *see* PIG

Boatbill A heron-like bird living in the swampy forests of South America. Its most remarkable feature is a wide, flat bill which is used to scoop small animals from the water.

Bobcat The wildcat of America, ranging from southern Canada to Mexico. It is actually a small lynx, about one metre long including its short tail. The

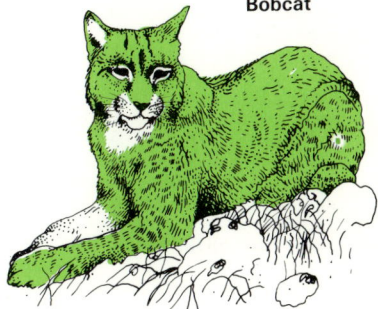

Bobcat

bobcat hunts mainly at night and it is an excellent climber. It feeds mainly on rabbits (cottontails) and small rodents.

Bog A plant community dominated by sphagnum mosses and growing on wet peat. Bogs develop in regions of poor drainage and where the rainfall is very high.

Bogbean (or buckbean) A marsh herb of Europe, northern Asia and North America. It has large, clover-like leaves and spikes of frilly pink and white flowers.

Bolete The boletes are a group of fleshy fungi in which the underside of the cap is sponge-like with tiny pores. The pores are the openings of closely packed tubes in which spores are produced. The penny bun bolete or cep is an edible and excellent fungus which is used in the manufacture of many mushroom soups.

Bolete (devil's boletus)

Bollworm This name is applied to the caterpillars of several moths which attack the bolls or seed-pods of cotton. Bollworms are widespread in the cotton-growing regions of the world and do a great deal of damage to cotton and also to many other crop plants.

Bone The skeletal material of vertebrates. Bone is living tissue permeated by minute canals and blood vessels. It is hard and able to stand up to all kinds of stress. The mineral which gives bone its hardness and its resistance to decay is a complex calcium salt which contains both phosphates and carbonates.

Bonito This name is given to several species of mackerel-like fish. The common bonito, found on both sides of the Atlantic in the tropical and subtropical zones, is a streamlined steel-blue fish about 60 cm long. Similar species occur in other parts of the oceans.

Bontebok An antelope closely related to the blesbok. The white face, white legs and white rump combine with the dark brown back to give the animal a very attractive appearance, and the name bontebok actually means 'painted buck'. The bontebok was once common in southern Africa, but is now a rare animal.

Booby A group of birds belonging to the gannet family found in tropical regions. They have goose-sized bodies with thick necks and large heads. Boobies breed on coasts and islands and, like gannets, they feed by diving for fishes out at sea.

Booklouse The name applied to a number of tiny plump-bodied wingless insects which feed on the paste and glue of book-bindings. They can become serious pests in libraries. Many related insects, some of them with wings, feed on the dust and pollen grains on tree-trunks.

Boomslang One of the rear-fanged snakes, in which some of the back teeth carry venom. Boomslangs are African snakes and they live in the trees, feeding on lizards, birds and small mammals. They average one metre in length and are usually green or brown in colour. Their venom is extremely poisonous.

Bootlace worm, *see* RIBBONWORM

Borage A low-growing herb of Europe and North Africa, whose hairy leaves have a cucumber-like flavour. It is used in salads and drinks. Borage has blue star-like flowers which attract lots of bees and was once specially planted by bee-keepers.

Bottlebrush plant An Australian plant whose scarlet flower spikes resemble bottlebrushes. They are greenhouse plants in cooler climates.

Bottle gentians, *see* GENTIAN

Bottlenose dolphin A favourite attraction of marine aquaria, the bottlenose grows to more than 3 metres in length and has a marked snout or 'beak'.

These dolphins communicate with each other by an elaborate system of clicks and squeaks. They are particularly common off the east coast of the United States.

Bottle-nosed whale, *see* BEAKED WHALE

Bottle-tree An Australian tree whose trunk is shaped like a bottle. There are two varieties, one with narrow leaves and the other with wide leaves.

Bougainvillaea Any of a group of South American shrubby climbing plants. Bougainvillaeas have small, insignificant flowers surrounded by brilliant red or purple BRACTS.

Bower bird There are several species of bower birds living in Australia and New Guinea. Some of them are beautifully coloured, but their most fascinating feature is the bower or courtship house which is built by the male. The bower usually consists of a platform of twigs and grasses partly or completely roofed over. The male normally decorates it with attractive pebbles, flowers, and other bright objects, and then he entices the female into it. Courtship and mating take place in the bower, but the female then makes a nest and lays her eggs elsewhere.

Bower bird

Bowfin This freshwater fish of North America is a 'living fossil' with many primitive features. The head is covered with bony plates and the rest of the body bears thin scales. A soft dorsal fin runs most of the length of the body, which may be up to one metre long, and the body ends in a rounded tail. The bowfin lives in sluggish waters and can breathe air with its lung-like swim bladder. It can live out of water for up to 24 hours.

Box A slow-growing evergreen tree of Europe, Asia and North Africa. It has small glossy leaves and produces hard, heavy, close-grained timber.

Boxfish, *see* TRUNKFISH

Brachiopod One of a group of bivalve marine animals. The name comes from a pair of food-gathering spiral arms (brachia). Brachiopods, also known as lamp-shells, resemble molluscs in appearance but there are sufficient differences for these animals to be placed in a phylum of their own, the Brachiopoda.

Bracken A species of FERN, found world-wide. It is one of the first plants to grow on burned or newly-cleared ground, and spreads rapidly by means of underground stems.

Bract A small leaf in the axil of which a flower stalk develops.

Bracteole A small leaf on a flower stalk.

Bracket fungus A leathery or woody fungus that grows like shelves on tree-trunks. Some species are edible though none is tasty.

Brahman cattle, *see* ZEBU

Brain The co-ordinating centre of the nervous system. It is the front part of the central nervous system and as well as its general co-ordinating function it is specially concerned with the special sense organs of the head. The brain is most highly developed in cephalopods and in vertebrates.

Bramble, *see* BLACKBERRY

Branchial Concerning the gills — e.g. branchial arteries.

Brazil nut (or para nut) The seed of a large evergreen tree growing in Brazil, the Guianas and Venezuela. The tree grows up to 36 metres. The spherical fruits each contain about 20 triangular seeds.

Breadfruit tree

Breadfruit tree A tree of the Pacific Islands up to 15 metres tall. Its large fruits have a mealy pulp similar in appearance to bread.

Bream A silvery fish living in slow moving streams and lakes throughout most of Europe. It belongs to the carp family and reaches a length of about 60 cm.

Brine shrimp A primitive relative of the crabs and lobsters which likes very salty water and is found in many parts of the world. Brine shrimps are about 1 cm long with two pairs of antennae, two compound eyes on stalks and a third small eye in the middle of the head. They swim upside down with their underside to the light.

Bristlecone pine The tree known to have the longest life-span. It is found in the dry mountains of California and Nevada. It has prickles on its cones, gnarled, twisted branches, and is an evergreen. The oldest living specimen is known to be about 4600 years old. It is never more than 9 metres tall.

Bristlemouth One of the commonest fishes in the ocean, the bristlemouth lives in deep water where it preys upon small animals. It has small eyes and rows of light-organs (photophores) on its flanks.

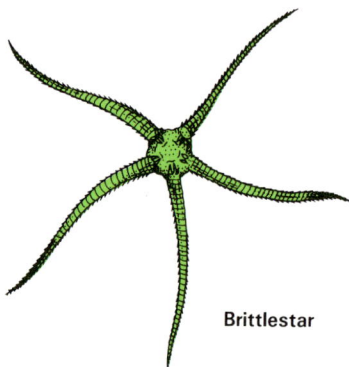

Brittlestar

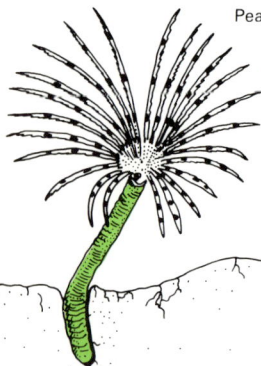

Peacock worm

Ragworm

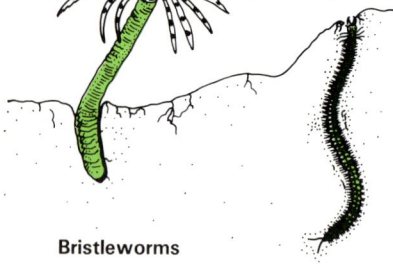

Bristleworms

Bristletail These wingless insects are found all over the world, the best known being the silverfish, often found in kitchens or pantries, or among books. They are up to 2.5 cm long and get their name from their three slender bristle-like 'tails'. The silverfish will eat anything but rarely becomes a pest.

Bristleworm This is the name given to a large group of marine worms which possess a number of prominent bristles. Examples include the ragworms, lugworms and fan worms. The latter animals use a fan of bristles to trap food particles in the water.

Brittlestar A distant relative of the starfish found in seas all over the world. It usually has five spiny arms (sometimes more) joined to a central button-shaped body. These arms break off easily as the name implies, but can be regrown. The bodies of brittlestars range in size from less than 1 mm to 10 cm across and the arms may span 60 cm. Along the underside of the arms are rows of tube feet which are used for feeding.

Broccoli A vegetable of the cabbage family, closely related to the cauliflower. It has thick clusters of green or purple flower buds, the edible part.

Bromeliad Any of a large group of American plants. They have rosettes of fleshy leaves which can absorb water and food. Many have no real roots and grow perched high up on forest trees. The PINEAPPLE is a bromeliad.

Bronchiole One of the branches of a bronchus within the lung.

Bronchus (plural bronchi). The vertebrate trachea divides into two bronchi, one leading to each LUNG.

Broom A shrub of Europe, Asia and Africa, with about 100 species. Brooms grow about 2 metres tall. They have small, often spiky leaves and produce seeds in pods. Most have yellow flowers.

Brown algae A large group of algae (singular alga) containing brown pigment. They all live in the sea and include the largest seaweeds — the kelps which grow to over 50 metres in length.

Brown bear The brown bear is found over a large area of the northern hemisphere and there are several subspecies. These include the Kodiak, Syrian, and grizzly bears. They vary in size, but all are large omnivorous animals weighing up to 750 kg. They have poor sight and rely on their senses of smell and hearing which are acute.

Brown hare Found over most of Europe with the exception of Ireland and northern Scandinavia, the brown hare is fond of open spaces. It has very long back legs and can run at speeds of

Brussels sprouts Vegetables of the cabbage family. Each plant has a tall stalk on which the sprouts, like miniature cabbages, form. The tops are leafy. Sprouts are food-filled buds.

Bryde's whale, *see* RORQUAL

Bryony The common name of two unrelated perennial climbing herbs. White bryony, found in Europe, western Asia and North Africa, is related to the cucumber and has greenish-white flowers. It climbs with tendrils. Black bryony occurs in the same areas. It is related to yams and has small greenish blooms. It twines around the other plants.

Bryophyte Any of a group of plants (Bryophyta) containing the mosses and liverworts.

Bryozoan, *see* MOSS ANIMAL

Buccal cavity The mouth cavity which in vertebrates contains the teeth and tongue.

Buckbean, *see* BOGBEAN

Buck-eyes, *see* HORSE CHESTNUT

Buckthorn A spiny shrub of Europe, North Africa and western Asia. It has oval leaves and small greenish flowers, and is used for hedging.

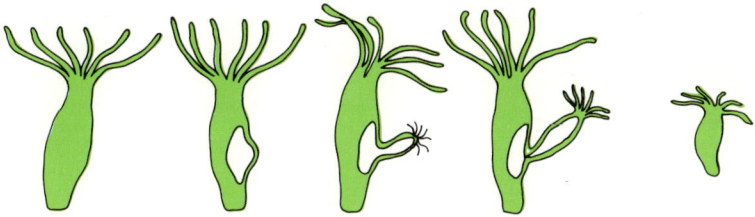

Hydra reproducing by budding

more than 80 km/h. These hares spend most of the day resting in hollows called forms and go out to graze at night. Similar species are found in other parts of the world.

Brush-tail opossum Also known as the vulpine, or fox-like opossum, the brush-tail opossum is found in Australia, Tasmania and New Zealand. It is about 60 cm long and its fur is thick and woolly, ranging in colour from silver grey to black. Its tail is prehensile at the tip.

Buckwheat A crop plant related to docks, grown for its seeds. It grows about 90 cm tall. Its seeds are like small beech nuts. The leaves are heart-shaped.

Bud A shoot, leaf or flower which has yet to open up.

Budding A form of PROPAGATION. Also a form of asexual reproduction in which detachable parts are formed on an organism and then released, forming a new individual. The division of yeast cells into two is called budding.

Buddleia A group of 100 species of deciduous and evergreen shrubs. They are native to the warmer parts of the world, especially eastern Asia. They are grown for their showy purple or orange flowers, which often form long spikes. The flowers of the buddleia attract so many butterflies that it is sometimes known as the butterfly bush.

Budgerigar The commonest member of the parrot family in Australia, the budgerigar travels in huge flocks. Being a seed-eating bird, it is a pest in grain-growing areas. The wild bird is bright green with black, yellow and blue markings but many different colour varieties have been produced by bird fanciers.

Budgerigars

Buffalo, North American, *see* BISON
Buffalo, Water, *see* INDIAN BUFFALO
Bug A term popularly used for most insects but scientifically applied to any member of the insect order Hemiptera, sometimes known as the 'true bugs' to avoid any confusion. Bugs all possess piercing mouthparts suited for sucking the juices of plants and animals but they differ considerably in appearance. The group includes, among many others, APHIDS, ASSASSIN BUGS, BACKSWIMMERS, BED-BUGS, CICADAS and POND SKATERS.

Bulb An underground storage and reproductive structure. It is a short stem surrounded by fleshy leaves. The flowering shoots develop from buds in the axils of these leaves. After flowering, a new bud (or buds) swells up as food accumulates in its leaves and this forms the new bulb.

Bulbul A group of thrush-like songbirds found in Africa, Asia and Australia. Some species are familiar garden birds — the equivalent of the European blackbird and sparrow.

Bulldog bat, *see* FISH-EATING BAT

Bullfinch A brightly coloured garden and woodland bird of Europe and Asia easily identified by its black and white markings and reddish underparts, especially bright in the male. It feeds on buds and seeds and is a serious orchard pest.

Bullfrog The bullfrog is a large North American frog up to 20 cm long, which is rarely found away from water. It feeds on a variety of insects caught at the surface of the pond but will also eat other animals, including small snakes and mammals.

Bullhead Several fishes are known by this name, but the true bullhead is a small, spiny fish found in clear streams over much of Europe. It is about 10 cm long, and has a broad head. The bullhead is also called the miller's thumb, while its relatives in North America are called catfishes.

Bulrush A tall water or marsh plant with strap-shaped leaves. Also known as reed-mace. The flowers form dense spikes at the top of the stems and later give rise to the familiar velvety brown clubs. The name bulrush is also

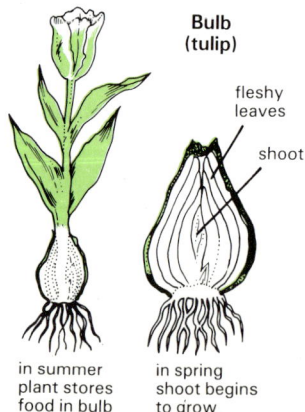

Bulb (tulip)

fleshy leaves

shoot

in summer plant stores food in bulb

in spring shoot begins to grow

Will Warhol hit £151m record?

COMPARED with the Mona Lisa as one of the world's greatest paintings, this portrait of Marilyn Monroe by Andy Warhol is predicted to fetch £151million at auction.

Such a price would make the 1964 pop-art piece the most expensive 20th century artwork to go under the hammer.

Christie's says Shot Sage Blue Marilyn – based on a promotional photo for the Hollywood star's film Niagara – is 'the most significant 20th century painting to come to auction in a generation'.

It is being sold in New York in May by a children's charity, the Thomas and Doris Ammann Foundation Zurich. Alex Rotter, of Christie's, said: 'Standing alongside Botticelli's Birth of Venus, Da Vinci's Mona Lisa and Picasso's Les Demoiselles d'Avignon, Warhol's Marilyn is categorically one of the greatest paintings of all time.'

at wedding of his daughter Alexandra, ce

applied to a tall, dark stemmed member of the sedge family which grows in still and slow-moving water.

Bumble-bee A large bee whose lazy, noisy flight from flower to flower is a familiar summer sight in many parts of the world. The body of a bumble-bee is covered in stiff hairs, often black and yellow, which pick up pollen from the flowers. Bumble-bees build nests from grass and leaves in cavities such as old mouse burrows.

Burchell's zebra, *see* ZEBRA

Burdock A coarse hairy weed producing thistle-like flowers. The flower-heads (burs) are hooked and cling to the coats of animals when the seeds are ripe. Seeds are scattered from the burs as the animals walk about. The big lower leaves are heart-shaped, and are eaten as a vegetable in Japan.

Burnet moth A group of brightly-coloured, day-flying moths that are most abundant in the Mediterranean area. Most have metallic blue or greenish-black forewings with scarlet markings.

Burrowing owl A small owl living over a wide area in the Americas. It hunts by both day and night, feeding mainly on rodents and insects. The bird gets its name from its habit of nesting in burrows on the ground, but it does not normally dig its own burrow. More often it takes over the deserted hole of a prairie dog or a fox.

Burrowing toad A small, oval-shaped Mexican toad which digs holes in which to escape the fierce heat of day. The burrowing toad inflates its body when alarmed to look half as large again as it really is.

Burying beetle This is the name given to various beetles which feed on the corpses of small animals. The beetles usually work in pairs — one male and one female — to bury a carcass by dragging out soil from under it. The beetles feed on carcasses and also lay their eggs on them. Burying beetles are also known as sexton beetles.

Bushbaby The four species of bushbaby, also called galagos, are relatives of the lemurs. All live in the African bush or scrub regions and they move around in the trees at night, feeding on insects, fruit, and birds' eggs. Bushbabies are furry creatures with large eyes and ears and long furry tails. The largest bushbaby is about 75 cm long including the tail.

Bushbuck A browsing antelope found in the forest and bush throughout most of Africa. The bushbuck stands about 75 cm high at the shoulder. It has a brown coat with white markings and the male has a bushy mane along the back. The sharp horns may reach over 50 cm in length, but females only rarely bear them.

Bush cricket A group of insects related to grasshoppers. Bush crickets have longer antennae than the true grass-hoppers and are sometimes (wrongly) called long-horned grasshoppers. Bush crickets 'sing' by rubbing their wings together and they feed mainly on other insects.

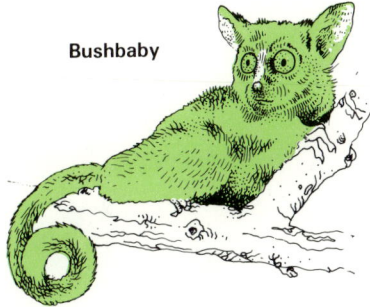

Bushbaby

Bush dog A South American dog only distantly related to the domestic dog but behaving very like it when tamed. It stands about 40 cm high at the shoulder, but its legs are relatively short. It is brown in colour and, unlike most animals, it is paler on top than below. It ranges from Panama to Paraguay and hunts at night in small packs.

Bushmaster, *see* PIT-VIPER

Bushpig The bushpig, also known as the red river hog, is a stout-bodied pig found over most of Africa. The head is large in proportion to the body and it bears tufted ears. The adults are reddish-brown or black and weigh up to 140 kg. Their main food is roots and fruits.

Bustard A group of large birds with stout legs and strong toes. Most species live in Africa, mainly on the ground in open country. The birds run well, but they can also fly in spite of their size. The great bustard is as large as a turkey, weighing up to 14 kg and having a wingspan of 2.5 metres.

Butcher bird The name given to various birds, including shrikes, which impale their prey on thorns.

Buttercup A perennial yellow flower flourishing in the northern hemisphere. The leaf shape gives the plant its other name, crowfoot. There are several species. Meadow buttercup grows up to 1.2 metres tall; creeping buttercup spreads by runners.

Butterfish An eel-like fish living between the tidemarks on both sides of the North Atlantic. The butterfish is about 15 cm long and it is brownish-green with darker markings. The name refers to the slippery nature of the fish. It is actually a kind of blenny.

Butterfly A large group of insects belonging to the order Lepidoptera which also includes the moths. Butterflies are generally day-flying creatures and moths night-flying, but there are many exceptions to this rule. A better way of telling the two apart is by the shape of the antennae. Butterflies have club-tipped antennae; the antennae of moths have many shapes but they are never club-tipped. Butterflies and moths feed on flower nectar and other liquids. Their mouthparts are extended to form a long tube or proboscis through which they suck up the liquid. When not in use the proboscis is coiled beneath the animal's head. Butterflies and moths pass through four stages in their life cycle: egg, larva (caterpillar), pupa and finally the adult butterfly.

Buttonquail Small ground-living birds of dry habitats, widely distributed from Spain through Africa and southern Asia to Australia. The female, unusually, is more brightly coloured than the male and attracts a partner in the breeding season with a strange booming call.

Butterwort An insect-eating plant of northern bogs and marshes. It has a rosette of sticky leaves which curl to trap any insect that lands on them.

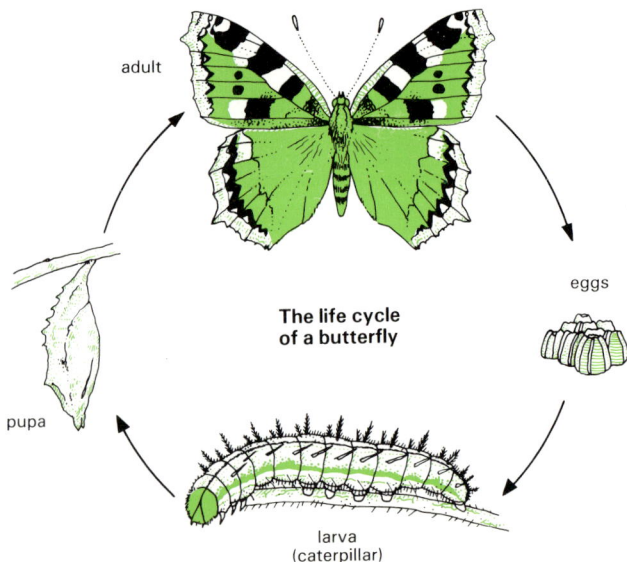

adult

eggs

The life cycle of a butterfly

pupa

larva (caterpillar)

Buzzard

Buzzard A large bird of prey found over a large area of Europe, Asia and Africa. It inhabits open and lightly wooded country, especially hilly regions, and soars high in the air on its broad wings. It feeds on rabbits and other mammals. The name is also used in America for the New World vultures, especially the turkey vulture of the United States.

C

Cabbage An important green vegetable with many varieties. Wild or sea cabbage, the parent stock, grows in Britain and northern Europe. The three main varieties of cabbage grown for food are white, Savoy and red. Brussels sprouts, broccoli, and cauliflower are also descendants of the wild cabbage.

Cabbage white butterfly, see WHITE BUTTERFLY

Cacomistle An omnivorous relative of the raccoon, found in the southwestern US and Mexico. The cacomistle has been given a variety of names including ringtail, raccoon fox and catsquirrel. It is a small furry animal with a fox-like face and a long squirrel-like

bushy tail ringed with black. Its fur is valuable and is sold as 'civet cat' and 'California mink'.

Cactus A flowering plant adapted to desert conditions. Many cacti have vast spreading root systems covering large areas close to the surface of the ground, and able to absorb great volumes of water quickly when the rain comes. The stems are distended and fleshy, sometimes forming enormous water stores. Often they are pleated, which allows them to expand like a concertina, thus increasing their water-carrying capacity. Cacti are distinguished from all other succulents by the possession of areoles. These are tiny 'pincushions' from which spines and glochids (barbed hairs) project. The original home of the cactus family was in America but many species are now found in other parts of the world.

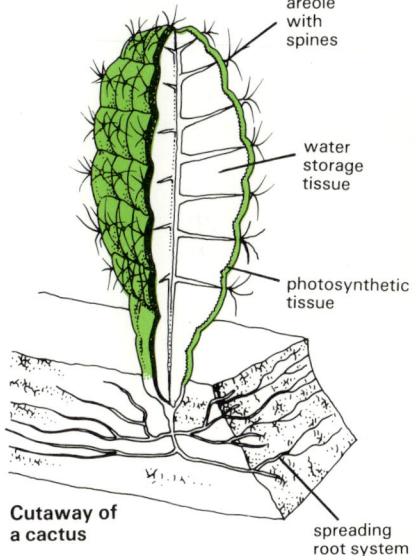

areole with spines

water storage tissue

photosynthetic tissue

Cutaway of a cactus

spreading root system

Caddis fly The name given to the insect order Trichoptera, of which there are some 5000 species throughout the world. The adult insects look rather like moths and fly mainly at night. Most of the larvae live in fresh water and breathe by external gills. Many of the larvae build themselves tubular 'houses' of stones, small shells, or leaf fragments.

Caecilian A limbless amphibian with a long cylindrical body. The animals vary in size from 20 cm to 1.3 metres and are usually a blackish colour. Their eyes are small and generally useless, but they have a sensory tentacle on each side of the head. Caecilians live underground in tropical regions of Central and South America, Southern Asia and parts of Africa.

Caecum A blind sac in the mammalian alimentary canal. It is VESTIGIAL in many species but is important as the site of cellulose digestion in some herbivores. The APPENDIX is at the end of the caecum.

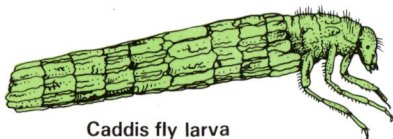

Caddis fly larva

Caiman A close relative of the alligator, differing mainly in having the skin of the underside covered in bony plates. There are five species, the smallest being the dwarf caiman up to one metre long, the largest the black caiman up to 4.5 metres long. They are found in the north of South America and southern Mexico.

Cake urchin, *see* SAND DOLLAR

Calabash A GOURD, growing on a tree or a vine of tropical America. People use the hard shells of the gourds as bottles, cups, and even cooking pots.

Calcicole Lime-loving. Plants of this type are found almost exclusively on calcareous soils. Examples include salad burnet, horseshoe vetch and many orchids.

Calcifuge Lime-hating. Plants of this type are rarely found where there is free calcium carbonate in the soil. They are typical of sandy soils. Examples include heather and rhododendron.

Californian sea lion, *see* SEA LION

Calyx The outer part of a flower, i.e. the sepals. They protect the inner parts of the flower when it is in bud and are usually green but sometimes they are brightly coloured as in the marsh marigold.

Camberwell beauty Known in North America as the mourning cloak, this butterfly is widely distributed over the northern hemisphere, though it is rare in Britain. The Camberwell beauty is chocolate-brown with a band of blue spots and a pale border along each wing. It hibernates in the winter. The caterpillars feed on the foliage of willow and birch.

Cambium tissue consists of actively dividing cells in a plant. It is found in VASCULAR BUNDLES and at the growing points of stems and roots. In woody plants it forms a ring just under the bark and produces a new ring (ANNUAL RING) of wood each year.

Camel There are two species of camel: the Arabian or one-humped and the bactrian or two-humped. They are desert animals, although only the bactrian camel still survives in the wild. A dromedary is a special breed of the one-humped camel used for riding. Both species are remarkably well adapted for desert life. Their humps consist largely of fat, which can be used to provide both energy and water when food is scarce.

Canada goose

male

Capercaillies

female

Camellia An evergreen tree of China and Japan, grown elsewhere for its showy flowers. The 80 species have dark fleshy leaves. One species is the TEA plant.

Campanula, *see* BELLFLOWER

Camphor tree A kind of laurel native to China, Taiwan and Japan. It is a tall tree with white flowers. When steamed its wood yields camphor, used in medicine and industry.

Campion A small wild flower of Britain and northern Europe, having white, red or pink blooms. There are several kinds.

Canada goose This large North American goose, now well established in Britain, is easily recognized by its black neck and head with a broad white chin-strap. Canada geese are equally at home grazing in a cornfield as in coastal marshes.

Canadian lynx, *see* LYNX

Canary A finch coming originally from the Canary Islands. The wild bird is brownish-yellow, but many yellow varieties have been produced in captivity. The canary is a very popular cage bird, with a sweet song. Like all finches it is a seed-eater.

Candytuft Any of several flowering plants of the mustard family, mostly low-growing. There are perennial and annual varieties, producing tufts of white, pink or red flowers, often with very sweet scents.

Cane sugar, *see* SUCROSE

Canine, *see* TOOTH

Canterbury bell, *see* BELLFLOWER

Cape buffalo Reputed to be the most dangerous of the African big-game animals, this bulky ox-like creature stands some 1.5 metres high at the shoulder and adult males may weigh up to one tonne. It has large horns which sweep first downwards and then upwards. Cape buffaloes live throughout Africa south of the Sahara. They are not normally dangerous unless they are unexpectedly disturbed.

Cape hunting dog A ferocious carnivore, only distantly related to the domestic dog, that ranges over most of Africa. It stands about 60 cm at the shoulder and has a mottled black, yellow and white coat with large, round ears. The animals hunt in packs and run down the antelopes on which they feed.

Capercaillie The largest member of the grouse family, the capercaillie is a bird of the northern coniferous forests. It is found over much of northern Europe and Asia. The male is a little

Cape hunting dog

Capers

under one metre in length and his plumage is predominantly black or dark grey. He has a fine tail, which he fans out and displays to the brown hen in the breeding season.

Capers The flower-buds of a southern European shrub, the caper bush. They are pickled to make a sauce. The sprawling plant has white or pinkish flowers.

Capillary Tiny blood vessel whose wall is only one cell thick. Capillaries arise from the repeated branching of arteries and run through all the tissues of the body, later joining up to form veins.

Capitulum An inflorescence typical of flowers of the family Compositae. It carries large numbers of tiny flowers called FLORETS.

Capsicum A perennial woody plant related to the tomato, native to Latin America. It has long pointed leaves. The fruits of some species, known as peppers, are used as food. The spices cayenne and paprika are made from two of the many other species.

Capsule A type of dry, dehiscent fruit formed from two or more carpels and opening in a variety of ways.

Capuchin The capuchins are small South American monkeys. The body is about 30 cm long, but the prehensile tail may be nearly twice as long again. The monkeys live in troops and feed mainly on fruit and insects. Capuchins are the traditional organ grinders' monkeys and they are exceptionally intelligent.

Capybara Also called the water cavy, the capybara is the world's largest rodent. It may reach well over one metre in length and looks like an enormous guinea pig. The capybara is found over much of South America but never far from water. It swims very well and feeds mainly on aquatic plants.

Caracal Also called the desert lynx, the caracal is a medium sized cat. It has a short reddish-brown coat, long legs, and long black ears topped by very long tufts of hair. The caracal ranges over much of Africa and parts of Asia. It feeds on a variety of small mammals and will also catch birds in flight by leaping up at them.

Caracara A group of predatory birds related to falcons which are widely distributed throughout the Americas. Caracaras often chase and bully other birds until they release the food they are carrying. The caracaras then drop down and collect it.

The carbon cycle. Carbon dioxide is removed from the air by plants during photosynthesis and is used to make sugars and proteins. Both plants and animals release the gas during respiration, and it is also returned to the atmosphere when things decay or burn.

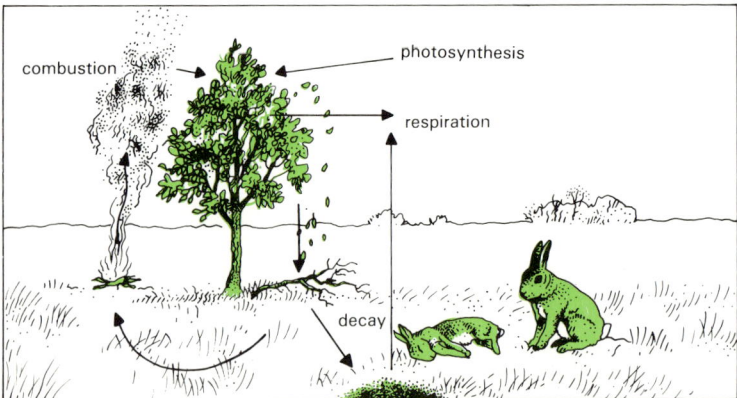

Carapace The hard protective shell which covers the head and thorax of many arthropods. It is made of a protein substance called chitin and usually has calcium compounds deposited on it to make it even harder. Carapace is also the name given to the upper part (half) of tortoise and turtle shells.

Caraway A herb of northern Europe and Asia, grown for its spicy seeds. It has a hollow stem and large, fern-like leaves. The small white flowers grow in large domed clusters (umbels).

Carbohydrate An organic compound containing carbon, hydrogen and oxygen. Examples are glucose (grape sugar), fructose (fruit sugar), sucrose (cane sugar), starch — an important storage material — and cellulose — the principal structural material in plants. Carbohydrates are the main fuels that are burned to supply the energy needed to power living processes.

Carbon cycle Carbon is an essential component of all living matter. Plants remove it from the air in the form of carbon dioxide and use it to make sugars, proteins and other materials. Animals obtain their carbon from the food they eat (derived ultimately from plants). But carbon is also returned to the atmosphere in many ways. Both plants and animals 'burn up' food materials to provide energy. This process (respiration) releases carbon dioxide back into the air. Bacterial decay of dead organisms also releases carbon dioxide. This interchange of carbon between the atmosphere and living organisms is called the carbon cycle.

Cardamom Any of several Indian plants, related to GINGER. It has lance-shaped leaves, small brownish flowers, and produces spicy seeds used as flavouring.

Cardinal A familiar songbird of North America, also known as the red bird. The male has a black 'bib', but otherwise he is bright scarlet. The female is more sombre, owing to a greater amount of brown in her feathers. Both sexes have a conspicuous crest and a stout, seed-cracking beak.

Caribou These animals are the reindeer of North America. They belong to the same species as the European reindeer, but the reindeer is usually

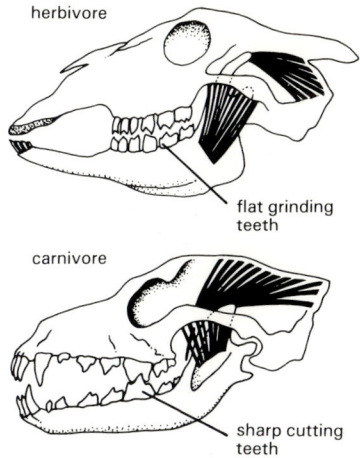

herbivore

flat grinding teeth

carnivore

sharp cutting teeth

Comparison between the skulls of a herbivore and a carnivore. The former is adapted for chewing with a sideways movement of the jaw, the latter eats with a shearing action and has sharp stabbing teeth.

regarded as a semi-domesticated race of the species. Caribou are up to 1.5 metres high at the shoulder and are generally brownish-grey in colour. They are unusual among deer in that both sexes have antlers. Caribou spend the summer on the open tundra, feeding on lichens and grasses. In the autumn most of them move back to the forests and spend the winter there feeding on lichens and browsing on the soft twigs of aspens.

Carnation A cultivated plant descended from various kinds of wild pinks — mainly *Dianthus caryophyllus* which is a native of southern Europe. Modern forms have many petals, unlike the native pinks with just five petals. The carnation has a strong, sweet scent.

Carnivore Any meat-eating animal, but the term is frequently restricted to mean members of the order Carnivora including the dogs, cats and weasels.

Carob An evergreen tree of the Mediterranean region also known as the locust tree. They have leathery

47

pods which are ground into pulp as feed for animals, and sometimes baked to form what is called locust bread.

Carp The common carp is a native of Japan, China and central Asia, but it has been introduced to many other regions. The carp prefers warm, shallow lakes and rivers with muddy bottoms, for it feeds on the mud-dwelling insects and worms. It is a favourite fish for ornamental pools and it can live for 50 years or more. The wild fish is olive green, but many colour varieties have been produced by breeders.

Carpel The female structure of a flower. There are often more than one and they may be joined (e.g. the plum has one carpel, the orange has several joined carpels, while the buttercup has many separate carpels). Each carpel contains one or more ovules that later become seeds. The carpel(s) develops into the fruit. Sometimes other parts of the flower contribute to fruit formation, and the fruit is then called a false fruit.

Carpet shark, *see* WOBBEGONG

Carragheen Also known as Irish moss, this is a small red tufted seaweed growing on the Atlantic coasts of Europe and North America. It produces a jelly-like substance used in food and medicines.

Carrion crow, *see* CROW

Carrot A long reddish root vegetable, descended from ancestors growing wild in Europe and Asia. It is a biennial plant with lacy leaves and is related to parsley.

Cartilage A skeletal tissue of vertebrates. It is not so hard or rigid as bone but is extremely tough and resistant to both compression and extension. Except in sharks, rays and a few other fishes whose skeletons are composed entirely of cartilage, it is found only in certain parts, such as the joints, of adult vertebrates.

Caryopsis The fruit of grasses. The ovary wall is joined to the seed coat, so that fruit and seed appear to be one and the same.

The tiger is the largest of the cat family. Though once common in the wild, its numbers have dwindled to a few hundred.

Cashew A small evergreen tree, native to tropical America. It has long, smooth leaves and fragrant rosy flowers. It is grown for the edible seeds inside its woody nuts. The trunk produces a gum used in varnish.

Cassava A South American shrub with a slender stem which can reach a height of 2.7 metres. Sweet cassava has roots eaten like potatoes; bitter cassava's roots are used to make tapioca.

Cassias, *see* SENNA

Cassowary A large flightless bird of the dense rain forests of New Guinea and northern Australia. It has coarse black plumage but the head and neck are naked and brightly coloured. Fruits and insects are the main foods.

Caste Social insects normally have a number of structurally and functionally different forms: honey-bees have queens, workers and drones; ants have various types of worker such as soldiers and foragers; termites have kings as well as queens and working forms. All the various forms are called castes.

Castor oil plant A shrub native to India and Africa. It is grown in many tropical lands for its seeds, which yield castor oil. It grows up to 12 metres tall, with smooth leaves up to 90 cm across. Apart from the oil, the plant is extremely poisonous and the leaves have been used as an insecticide.

Cat The present-day domestic cat is thought to be descended mainly from the African bush cat. Domestication took place in Egypt where the cat was regarded as sacred. Today there are many breeds of domestic cat. Closest to the cats of Ancient Egypt is the Abyssinian. All domestic cats retain the hunting instincts of their ancestors and go feral, or wild, with surprising ease.

Cat-bear The cat-bear, also called the red panda or lesser panda, is the nearest living relative of the giant panda. It lives in the Himalayas. Its fur is a rich chestnut brown, with white marks on the face and dark rings on the long bushy tail. Although classified as a carnivore, the cat-bear feeds almost entirely on fruit and leaves. It spends most of its time in the trees, sleeping by day and coming out to feed by night.

Catbird The name given to several unrelated birds. The Australian cat-birds are closely related to the BOWER BIRDS. The North American catbird is a member of the mockingbird family. It is a familiar garden bird throughout most of the United States, well known for the mewing notes of its song.

Caterpillar The larval stage of various insects (especially butterflies and moths). The soft body carries three pairs of jointed legs on the thorax and a number of stumpy legs behind.

Catfish A large group of mainly fresh-water fishes which have 3 pairs of 'whiskers' or barbels around their mouths. Most are scavengers living in slow-moving rivers in North and South America, Africa and Asia. They range in size from 3 cm to 3 metres in length.

Catkin A spike, often hanging, of simple flowers — e.g. hazel.

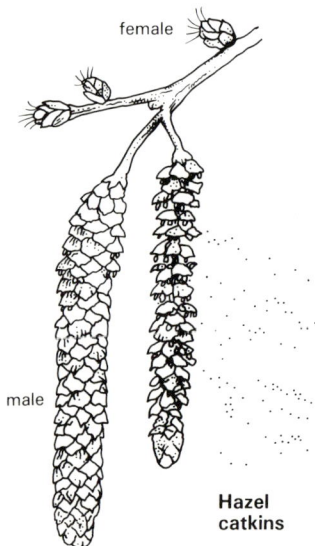

female

male

Hazel catkins

Cat's-ear A European perennial herb. It has dandelion-like flowers on 60 cm stalks bearing scattered scale-like bracts. The leaves are dull green. There are several species.

Cedar

Catshark A group of small sharks distinguished by their picturesque patterns of stripes and mottlings.

Cat-squirrel, *see* CACOMISTLE

Cattle Domestic cattle probably originated from the aurochs, an extinct species of ox. Many strains have been bred for particular purposes, such as meat or milk production, and to suit particular environments.

Cattle egret, *see* EGRET

Caudal Concerning the tail.

Cauliflower A kind of cabbage, similar to broccoli. It produces a dense head of white flower buds, packed with food, and it is these buds and their stalks that we eat. The flowers are yellow when they are allowed to open.

Cauline Of the stem. Cauline leaves are carried on the upper part of a stem but do not bear flower shoots in their axils.

Cave fishes A number of fishes spend their whole lives in underground caves and lakes. They have little or no pigment, and appear pinkish because the blood shows through the more or less scaleless skin. Most are blind and the largest is only 20 cm long. Cave fishes probably feed mainly on insects that fall into the water.

Cavy The name applied to a number of South American rodents of the genus *Cavia*, the best known being the domestic GUINEA-PIG which has been derived from the Peruvian cavy. The Patagonian cavy is a hare-like rodent found on the plains of Patagonia.

Cayenne pepper The dried and ground pod-like fruit of various species of capsicum pepper. The plant is a semi-woody many-branched perennial, native to tropical America.

Cedar A large coniferous evergreen tree of Asia, North Africa and Europe. It has wide spreading branches and whorls of stiff needles, and produces light, reddish-brown timber. The term cedar is used for the wood of several other trees.

Celery A biennial vegetable related to parsley. It is grown for its crisp leaf stalks. It has large compound leaves, and if not harvested after its first season produces heads of small white flowers on a long stem.

Cell All living organisms are composed of one or more cells — tiny compartments normally invisible to the naked eye — within which the vital processes of life go on. In protozoans the body consists of only one cell and all the processes have to go on in that cell. In many-celled animals, however, there are several different types of cell, each specialized for a different function. Examples are muscle-cells, nerve-cells, bone-cells, blood-cells and many others. The basis of all living cells is protoplasm, a very complicated mixture of organic and inorganic substances.

Cellulose A carbohydrate that forms the principal structural elements in plants, composing the cell walls. Its molecules are formed by the union of many glucose molecules, forming fibres. The latter are of great use to man as the basis of textiles.

Cementum, *see* TOOTH

Centipede

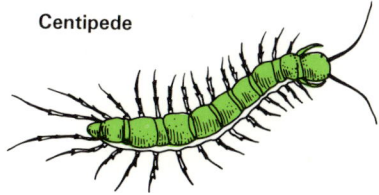

Centipede A group of arthropods which have a pair of walking legs on almost every segment. Although the word centipede means '100 feet', the creatures have varying numbers of feet — from 15 to 177 pairs. Centipedes live in damp places, particularly under logs and stones. They are active hunters, killing insects and other small creatures with their poisonous claws.

Central nervous system (CNS) The brain and spinal cord or main nerve cord of the body. The central nervous system co-ordinates the activity of the whole body.

Century plant, *see* AGAVE

Cep, *see* BOLETE

Cephalopod A member of the class of molluscs known as Cephalopoda, containing the squids, octopuses and cut-

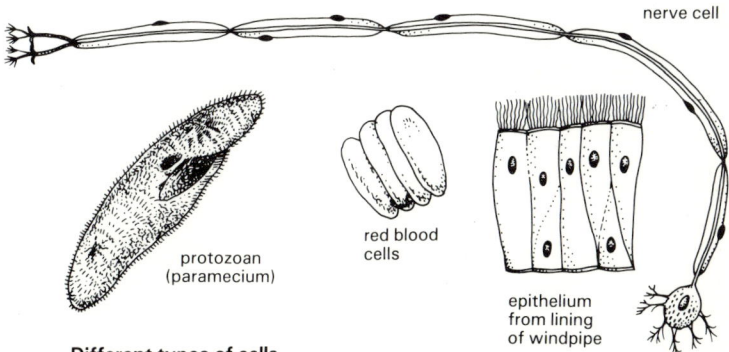

nerve cell

protozoan (paramecium)

red blood cells

epithelium from lining of windpipe

Different types of cells

Barley

Rye

Wheat

Oats

Maize

Rice

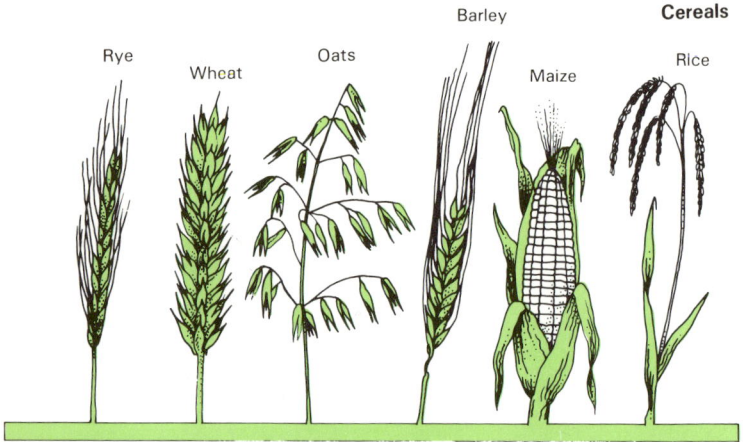

tlefish, all of which are marine. There are many extinct members — ammonites, belemnites and others — typically with a chambered shell. Modern forms, however, except for the pearly nautilus, have a very reduced shell. In size they vary from a few centimetres to 18 metres — the length of a giant squid including tentacles. Cephalopods have a large distinct head marked off from the rest of the body by a narrower 'neck' and a pair of prominent eyes. Eight or ten muscular arms surround the head. They are equipped with numerous suckers which provide the animals with an extremely strong grip. The mouth is armed with a horny beak made of chitin and the tongue is covered with rows of curved rasping teeth (radula). Compared with other molluscs, cephalopods lead a very active life and have a correspondingly better developed nervous system with a large brain. The animals feed on crabs, prawns, fish and other molluscs. Squids swim freely, others tend to rest on the seabed, though the cuttlefish can swim gently by flapping 'fins'.

Cereal A member of the grass family (Gramineae) whose seeds are used for food. Examples include rye, oats, wheat, maize and barley.

Cervical Concerning the neck — e.g. cervical vertebrae.

Cestode A member of the class Cestoda. Cestodes are flatworms, all of which are parasitic in the food canals of other animals. They are commonly known as TAPEWORMS.

Cetacean A member of the aquatic mammalian order Cetacea which includes the whales.

Chaeta (plural chaetae). A bristle, especially of annelid worms.

Chaffinch One of the commonest of European birds, the chaffinch is easily distinguished by its handsome plumage. The wings and tail are black with prominent white bars. The male has a pink face and underparts and his crown is blue, while the female has a greyish-brown head and breast.

Chalcid wasp A minute relative of bees and wasps. The chalcids are mostly parasites of other insects.

Chameleon A group of lizards which are famed for their ability to change colour. They do this by altering the distribution of pigment in their skins. Other unusual features include eyes which can be swivelled around independently, and a long tongue which is catapulted out to catch insects. Chameleons are rather flattened from side to side and they are often decorated with brilliant colours, 'warts' and horns. Most species live in Africa and range from 5–60 cm in length.

Chamois A goat-like mountain animal found in various parts of Europe and Asia Minor. It is a sturdy creature about 1.25 metres long and weighing up to 40 kg. Both sexes have sharply pointed horns. The long coat is tawny in summer and darker in winter. The animals live mainly in alpine forests and feed on a variety of plants.

Chamomile Any of a group of strongly scented small herbs with daisy-like flowers. It has finely-divided leaves. This European plant is used to make medical preparations.

ner of the mouth. Over short distances it is probably the fastest land animal in the world, with a speed approaching 100 km/h. It catches antelopes by a single leap or after a short sprint. The cheetah once ranged from India to South Africa, but its main stronghold is now in East Africa.

Chela The claw or pincer of crabs, lobsters and some other arthropods.

Cherry Any of several kinds of deciduous trees of temperate lands. They have oval leaves and black or red stone-fruit. Some Japanese varieties

Chameleon

Chanterelle, *see* AGARIC

Characin A large family of tropical freshwater fishes living in Africa and the Americas. Its members range from the brilliantly coloured TETRAS of aquarium fame to the ferocious PIRANHA.

Charlock An annual and a member of the cabbage family, charlock is a cornfield weed of Europe, North Africa and western Asia. It is also called wild mustard. It has bristly, 60-cm stems and leaves and small yellow flowers. Its seeds can remain dormant for 50 years.

Cheetah A long-legged cat with a spotted coat and a characteristic black stripe running from the eye to the cor-

have no fruit and are grown for their pink blossom.

Cherry laurel An evergreen shrub of the eastern Mediterranean with shiny leaves and white flowers. Leaves, fruit and bark are rich in prussic acid and are poisonous.

Chervil A biennial herb of eastern Europe and western Asia, sometimes called sweet cicely. Its much divided leaves are used as flavouring.

Chestnut The sweet or Spanish chestnut is a deciduous tree of the northern hemisphere with wide-spreading branches. It has broad, toothed oval leaves and reddish edible nuts enclosed in spiny green cases. The male flowers are CATKINS. It reaches a height of 30 metres. The HORSE-CHESTNUT belongs to a different family.

Chevrotain Also known as mouse deer, the chevrotains are the smallest of the ruminant animals. They are rarely more than 30 cm high at the

Cheetah

Chickaree

shoulder. They are not true deer, and they lack antlers. There are four species, living in the forests of Africa and southern Asia. They eat mainly grass and leaves, but also take insects, fish, and small mammals.

Chickaree This is the red squirrel of North America. It is about two-thirds the size of the grey squirrel, and has a tawny brown coat. The eyes are ringed with white. The chickaree is found throughout coniferous and deciduous forests, and a related species is found along the west coast of the United States. The squirrels are active throughout the year and actually tunnel through the snow to find food.

Chicken The domestic chicken is derived mainly from the jungle fowl of southern Asia. There are numerous varieties including egg-laying, meat-producing and dual-purpose breeds.

Chickweed A low-growing annual herb of northern Europe. It has little oval leaves in pairs, and star-like flowers which even appear in winter.

Chickory A perennial herb of Europe, North Africa and India. It has long jagged leaves and bright blue daisy-like flowers. Its long roots are roasted and ground to add to coffee. A leafy variety is grown for salads, the basal leaf cluster or heart being called a chicon.

Chicon, *see* CHICORY

Chile pine Also known as the monkey puzzle tree, the Chile pine is an evergreen native to the western slopes of the South American Andes where it grows up to 30 metres. It is widely planted elsewhere as an ornamental tree because of its strange branching form.

Chillies The fruit pods of capsicum peppers, used as very hot spice.

Chimaera The chimaera or king herring is one of a small group of fishes which possess features characteristic of both the cartilaginous fishes (sharks and rays) and the bony fishes. They have cartilaginous skeletons, but have gill covers like those of bony fishes. The king herring has a large head, and the body then tapers back to the tail. It feeds mainly on bottom-dwelling molluscs and crustaceans, and reaches a length of 1.5 metres.

Chimney swift, *see* SPINETAIL

Chile pine

Chimpanzee One of the great apes and the nearest in intelligence to man, the chimpanzee lives in the rain forests of Africa and spends much of its time in the trees. Chimpanzees live in small bands, but there is little social structure and members come and go as they please. They eat mainly fruit and other vegetable food.

Chinchilla Best known for its remarkably soft grey fur, the chinchilla is a South American rodent. It is about 25 cm long and looks rather like a squirrel, although its long back legs are more like those of a rabbit. Chinchillas live in burrows in parts of Chile. They feed on coarse grasses and herbs.

Chinese lantern Any of about 100 Asian perennial herbs with woody stems. The fruit is enclosed in a lantern-like calyx, which is dried and used as a window decoration.

Chinese water deer A small deer about one metre long. The bucks do not have antlers but instead possess a pair of sharp tusk-like upper canine

54

teeth. It is found in the valleys of the Yangtze river in China, in Korea, and, since its introduction towards the end of the last century, the Chinese water deer can be found in certain areas of Great Britain.

Chipmunk A group of ground squirrels found in North America and Asia, where they live in open country and feed on berries, grasses, and a wide variety of small animals. They are great hoarders of food for the winter, although they do not actually hibernate. Chipmunks range from 10–25 cm in length, excluding the bushy tail.

Chital, *see* AXIS DEER

Chitin Horny, nitrogen-containing material which makes up most of the external skeleton (cuticle) of arthropods and also the chaetae of annelid worms.

Chiton One of a group of primitive molluscs, sometimes called coat-of-mail shells, from the eight overlapping plates which cover their backs. They graze on the algae coating rocky shores, and cling limpet-like to the rocks when the tide is out.

Chives A green vegetable of the onion family. It grows from a bulb and has long tubular leaves and globular heads of pale purple flowers. It is used as a flavouring.

Chlorophyll The green colouring matter (pigment) found in most plants, other than fungi, which gives them their green appearance. It plays a major role in PHOTOSYNTHESIS by trapping the energy of sunlight which is used for the manufacture of foods such as sugar.

Chloroplasts The small chlorophyll-containing bodies normally present in the leaf cells of green plants. Photosynthesis is carried on inside the chloroplasts.

Chordate A member of the phylum Chordata which includes all animals which at some stage of their life have gill slits and a supporting rod of elastic material (the notochord) and always have a hollow dorsal nerve chord (i.e. one which runs through the back). In fishes, amphibians, reptiles, birds and mammals, the notochord is replaced by a backbone early on during development. In reptiles, birds and mammals gill slits show at an early stage of development but later disappear. Most amphibians have gills only as tadpoles.

Chough The two species of chough are members of the crow family. They are a little larger than jackdaws and have glossy black coats. The common chough, found mainly by the sea,

Chimpanzees

Chrysalis

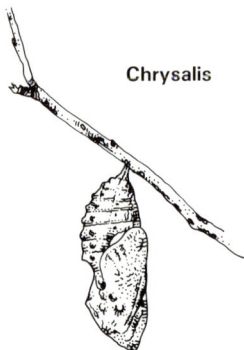

ranges in a narrow belt from Ireland to China. It has a long curved red beak. The alpine chough has a shorter yellowish beak and lives in the mountainous regions of Europe and Asia. Both species are scavenging birds.

Christmas rose A European evergreen herb of the buttercup family which produces its white flowers in winter. The similar lenten rose blooms in February-March. The plants are about 30 cm tall.

Chromosome One of the microscopic thread-like structures that occur in the cell nucleus and carry the GENES. They play a major role in HEREDITY. In the normal body cells of most plants and animals the chromosomes exist in pairs. This is the DIPLOID condition. The number of pairs varies with the species. Human body cells, for example, contain 23 pairs of chromosomes. When the reproductive cells or gametes — sperm and egg cells — are formed, the chromosome pairs split up and only one of each pair goes to each gamete. Human gametes thus contain just 23 chromosomes. This is the HAPLOID condition. At fertilization, when sperm and egg cell join, the diploid condition, with its full complement of 23 pairs of chromosomes, is regained. The members of each chromosome pair are normally identical in appearance and they carry genes affecting the same features in the body. They are known as homologous chromosomes. In many animals, however, there is one special pair of chromo-

somes which are not necessarily alike. They are known as SEX CHROMOSOMES because they control the sex of the animal. Human sex chromosomes are known as the X and Y chromosomes. Females have two X's in their body cells, while males have an X and a Y. (See also MEIOSIS; MITOSIS.)

Chrysalis The pupa or resting stage between larva and adult of butterflies and moths.

Chrysanthemum Any of about 100 species of perennial plants mostly native to China in which the stems are soft at first but become quite woody with age. Chrysanthemums have large composite flower-heads in which the florets are long and tubular and often ragged. Chrysanthemums are members of the daisy family.

Chuckwalla The chuckwalla is a plump lizard, usually about 30 cm long. It is a mottled, brownish-black creature and it lives in the hot desert regions of North America. Unlike most lizards, it eats plant material. When disturbed, the chuckwalla scuttles into a rock crevice and wedges itself there by inflating its body with air. It is then very difficult to dislodge.

Cicada A large group of bugs found mainly in the tropical regions. The males (and in some species the females too) 'sing' by rapidly vibrating little membranes on the sides of the body. This makes a very shrill whistling noise. The insects feed by sucking sap from trees. The young cicadas feed on the roots of trees and some species take 17 years to reach the adult state.

Cichlid The cichlid fishes, favourites with tropical fish keepers, live in rivers and lakes all over Africa and throughout most of South and Central America. The body is flattened from side to side and many species are brightly coloured.

Cilium (plural cilia) A very short fine 'hair' projecting from the surface of certain types of cell. There are always many and they beat with a definite rhythm. They are found in most animal groups and their function is to move the animal through the water (see PARAMECIUM) or to move liquids along inside the animal. The lining of the human nose and windpipe is covered with cilia which beat and sweep out mucus and any dust that it traps.

Cinchona Any of 38 species of shrubs and small trees from South America. The bark of several species yields the drug quinine. They have long oval leaves and clusters of pink or white blooms.

Cinnamon A small evergreen tree of the laurel family, native to Sri Lanka and Malabar. It grows up to 9 metres tall, and has oval leaves and acorn-like flowers. A spice is made from its bark.

Cinquefoil A small creeping herb of the rose family with five-petalled yellow flowers. Its leaves are on long stalks. Several species are found throughout the northern hemisphere.

Citron A thorny Indian tree of the orange family. It has lance-shaped leaves and purple and white flowers. The fruit is up to 23 cm long, with white flesh. Its thick rind is often candied.

Clam The name given to many bivalve molluscs. The largest is the giant clam of coral reefs. It may measure one metre across and weigh 250 kg. In North America the name is most commonly used for edible shellfish such as the round or hard-shelled clam, the soft-shelled clam and the long clam or sandgaper.

Class A category used in the CLASSIFICATION of living organisms.

Classification The division of living things into groups of related forms. Amongst the many different kinds of plants and animals some are more alike than others. The differences and likenesses are used to place them into groups which may be subdivided many times, the members of a sub-group having more features in common than the members of a group. A classification is necessary so that the relationships of the many kinds of plants and animals are understood. Besides tak-

Giant clam

Citrus Any of a group of evergreen trees, producing juicy fruit. They include CITRON, GRAPEFRUIT, LEMON, LIME, MANDARIN and ORANGE.

Civet A group of carnivorous mammals related to cats and hyenas. They have long narrow bodies and pointed muzzles, but their coats are cat-like. They live in Africa and Asia and keep mainly to the forests.

Cladode A stem that is leaf-like in appearance and function, as found in the butcher's broom and some cacti.

ing into account the plants and animals living today, any system of classification must include forms which are now extinct.

Each kind of plant and animal has two Latin names: the first is its generic name (the name of its genus), the second its specific name (the name of its species). Closely related species all belong to the same genus. Thus the different kinds of buttercup all belong to the genus *Ranunculus*. One species is called *Ranunculus repens*, another

Ranunculus acris and a third *Ranunculus bulbosus*. The Latin names may appear cumbersome but they are standard throughout the world. Common names not only vary locally, but from country to country and 'buttercup' may mean something completely different to a person from another area — if it means anything at all. Closely related genera are grouped into families. For example *Ranunculus, Clematis* (e.g. old man's beard), *Caltha* (e.g. marsh marigold) and *Anemone* are grouped in the family Ranunculaceae. Related families are arranged into orders. Thus the Ranunculaceae are in the order Ranales and this order, with many others, forms the class Dicotyledonae of the subdivision Angiospermae. The Angiosperms or flowering plants are a subdivision of the division Spermatophyta, the seed-bearing plants. The classification of *Ranunculus repens* may be written:-

Division:	Spermatophyta (= seed plants)
Subdivision:	Angiospermae (= flowering plants)
Class:	Dicotyledonae
Order:	Ranales
Family:	Ranunculaceae
Genus:	*Ranunculus*
Species:	*repens*

Latin names (genera and species) are always italicized.

The largest grouping within the animal kingdom is the phylum, and it is divided into a number of classes.

Clavicle A bone of the pectoral girdle. In mammals it is known as the collar bone.

Clawed frog This animal belongs to the group of tongueless frogs which use their front teeth to catch small fish and other aquatic creatures. The front legs are short and weak but the back legs are strong and carry large webbed feet. The three inner toes of each back foot have sharp claws. The clawed frog rarely leaves the water.

Cleaner fish, *see* BARBER FISH

Clematis A group of 250 woody climbing plants, many of which are grown for their beautiful flowers. The group also includes traveller's joy which smothers the hedgerows all over Europe. Some species are deciduous, others evergreen, and there are also some non-climbing species.

Clementine, *see* MANDARIN

Click beetle A group of narrow-bodied insects which get their name from their ability to flick themselves upright with a loud 'click' if they fall on their backs. Click beetle larvae are known as wireworms, and several species are serious garden pests.

Climax vegetation A plant community, the composition of which is more or less stable, and whose character is mainly determined by the prevailing climatic conditions. In British lowland conditions, for example, open country is gradually invaded by trees of various kinds until oak becomes the dominant plant species, and in equatorial regions it is dense evergreen rain forest. In slightly drier areas it is grass-land, and in cold areas of northern Europe it is coniferous forest.

Climbing perch A small fish found in southern Asia from India to the Philippines. Part of the gill cavity is separated off and functions as a lung. The fish can therefore live out of water for quite long periods. It can walk about on the land by propping itself up on its fins and spiny gill covers.

Cobra

Clothes moth The name given to several species of moth whose larvae feed on fur and wool. Clothes moths are native to Europe but have been introduced into other continents.

Clouded leopard Found in the forests of South-East Asia and the islands of Indonesia, clouded leopards are agile tree-climbing cats. They have extremely long tails accounting for almost half of their 2-metre length. The coat is pale brown with greyish patches enclosing small dark spots.

Clove An evergreen tropical tree which grows up to 9 metres in height and has purple flowers. The dried flower buds are used as a spice. Though native to Indonesia most of the spice production is done in Zanzibar and Malagasy.

Clover A low-growing herb with its leaves divided into three leaflets (rarely four). It is found wild in meadow-land, and is cultivated as a pasture crop and to put nitrogen back in the soil via its root nodules. Clover has dense heads of slender pea-like flowers.

Clown fish, *see* DAMSELFISH

Club moss Any of a primitive group of plants related to ferns and horsetails. Club mosses have stems clothed with slender green leaves. They reproduce by spores which are generally borne in cones. Most are tropical but some are found on European heaths and moorlands.

Coalfish A member of the cod family about one metre long. It gets its name from its dark back, but it is also known as the pollack. It lives in the North Atlantic and the Mediterranean and is fished commercially.

Coati A carnivorous mammal related to the raccoons found in the forests of South and Central America. The body is about 60 cm long and the banded tail is the same length again. The general colour is reddish-brown. Coatis are industrious little creatures, foraging nearly all the time for small animals. They will also eat fruit and they often climb trees to reach it.

Coat-of-mail shell, *see* CHITON

Cobnut, *see* HAZEL

Cobra A group of poisonous snakes found in Africa and Asia. They average about 2 metres in length and have a characteristic hood behind the head.

Cockatoo

The snake rears up and expands the hood when it is disturbed or excited. Cobras feed mainly on rodents, which they kill with a paralysing venom. The African spitting cobra can spit venom up to 2.5 metres. The largest species and one of the most deadly snakes in the world is the king cobra — up to 5.5 metres in length. It lives in India and the Far East.

Coca Any of a group of South American shrubby trees, with green, bitter-tasting leaves. The drug cocaine is extracted from the leaves.

Coccyx A rudimentary tail skeleton consisting of several fused vertebrae.

Cochlea That part of the ear which is concerned with the 'sorting out' of the sounds received.

Cockatoo The various species of cockatoo differ from the other parrots by having prominent crests on their heads. Most are white, but a few are black or grey. Cockatoos live in the Australasian region, often forming large bands. Some species are pests because they take a lot of grain and often attack fruit trees.

Cockchafer A large beetle with prominent fan-like antennae. Cockchafers have a habit of flying noisily into lighted windows and lamps on summer evenings. They are also known as May bugs. Adults and larvae are serious pests of trees and cereal crops respectively.

Cockle A common bivalve mollusc with a world-wide distribution. The two identical halves of the shell are ribbed and in some species have spines.

Cocklebur Any of a group of rough-leaved annual weeds, native to the Americas. The seeds are contained in spiny burs which cling to clothes and animals' coats.

Cock-of-the-rock Two uncommon birds of South America recognizable by their fan-like crests. Females are dull brown, but the males are brilliantly coloured. One species is orange and one is bright red.

Cockroach A large group of scavenging insects feeding on a variety of dead plant and animal material. Most of the 3,500 species of cockroach live in the tropics, but some of them have become established as pests in buildings in the cooler parts of the world.

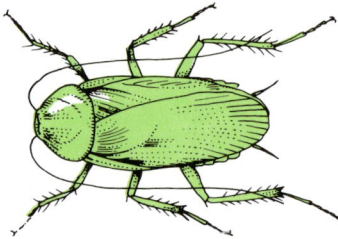

Cockroach

Cocoa A powder prepared from the seeds of the cacao tree. The tree is native of tropical America but is grown in many tropical areas, especially in Africa. The cacao tree is up to 12 metres tall and has wide branches, leathery leaves, and pink flowers growing directly on the trunk.

Coco de mer (or double coconut) A palm tree native to the Seychelles. Its thin stem is up to 30 metres tall. The tree has huge fruits up to 18 kg in weight and often found floating at sea.

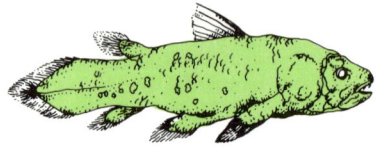

Coelacanth

Coconut palm A tall palm tree, native to the Pacific islands and South-East Asia but now growing throughout the tropical areas. It can reach 15 metres or more in height and has a crown of feathery leaves. The large seeds, inside a woody and fibrous fruit, hold a white 'meat' and a milky substance. Copra, the dried meat, yields a valuable oil.

Cocoon A protective covering. Earthworms and many spiders wrap their eggs in cocoons. Many caterpillars spin silken cocoons around themselves as a protection when they are about to pupate.

Cod A valuable food fish, the Atlantic cod lives on the continental shelves around the North Atlantic. It can reach almost 2 metres in length and weigh 100 kg, although it is usually a good deal smaller. During the summer the cod feeds mainly on other fishes, but it is more of a bottom-feeder during the winter.

Codlin moth The larva of this small, inconspicuous moth is a serious pest in apple orchards. It feeds inside the fruit, often entering by the 'eye' and leaving little outward sign of its presence.

Coelacanth A primitive fish believed to have been extinct for more than 70 million years until a specimen was caught by a South African fishing boat in 1938. Several other specimens have been caught since. The coelacanth has fleshy lobes on its fins and heavy scales. It represents a very early stage in the evolution of the bony fishes.

Coelenterate A member of the phylum Coelenterata composed of animals having a central body cavity with a single opening, the mouth, which is surrounded by a ring of tentacles, many armed with stinging cells. They include the PORTUGUESE-MAN-OF-WAR, HYDRA, JELLYFISHES, CORALS and SEA

ANEMONES. Many coelenterates possess two distinct forms in their life-history — a fixed, tubular polyp and a free-swimming, disc-shaped medusa — but some groups have either polyp or medusa only.

Coffee tree A shrubby, evergreen tree which originated in Ethiopia. It has glossy leaves and white flowers. The red berries contain the seeds or 'beans' which are roasted to make the drink coffee.

Colchicum Also known as meadow saffron, this is a poisonous plant of European meadows. It has pale purple crocus-like flowers in the autumn, but the glossy green leaves appear in spring. It yields the drug colchicine.

Coleoptile The sheath surrounding the young shoot of grass seedlings.

Collared dove Easily distinguished by a narrow black band on the back of the neck and black wing tips, the collared dove has spread across Europe from Asia during this century. It is equally at home in fields and gardens and is becoming an agricultural pest.

Collared dove

Collared peccary, *see* PECCARY

Colobus monkey A small group of monkeys which live in the dense forests of central Africa and rarely descend from the trees. They have long fur and reach up to 1.5 metres in length, including about one metre of tail. Colobus monkeys feed mainly on leaves.

Colorado beetle A native of North America, the Colorado beetle was quite harmless until potato growing reached its home. It then transferred its attention to potatoes and has since spread over large areas of America and Europe. By eating the potato leaves, it drastically reduces the yield of tubers. The beetle is nearly 1 cm long and has black and yellow stripes.

Columbine A group of perennial plants of the northern temperate zone belonging to the buttercup family. Columbines grow to 60 cm in height. The drooping flowers are of shades of pink or blue. Also known as aquilegia.

Comb jelly Small marine animals related to jellyfishes, the comb jellies are so named because of the rows of tiny hairs that resemble minute combs on the sides of their bodies. They are also called sea gooseberries due to the size and shape of most of the species. Comb jellies are found in all the oceans and are usually colourless and quite transparent. They swim about by waving their hairs or cilia.

Comma butterfly Widespread over southern Britain and Europe, this butterfly is easily recognized by the jagged shape of its wings. It is brownish-orange with dark markings on the upperside, while the underside is dark brown with a white comma-shaped mark which gives the insect its name. The comma butterfly hibernates during the winter in fairly exposed posi-

Colobus monkey

tions, such as hedges. With its wings folded it looks remarkably like a withered leaf. The caterpillars feed on nettles and elm foliage.

Community (of plants) is applied to any collection of plants which make up a distinct type of vegetation — from woodlands to the scant plant growth on the flanks of a sand-dune. A typical example is an oak-wood. In springtime, growing close to the ground are clumps of moss, white anemones, lesser celandines, winter aconites, and thick green spreads of dog's mercury; above is an undergrowth of woody shrubs such as hazel, hawthorn and sloe; higher still are branches of the oak trees themselves, supported on tough rigid trunks. All these plants are growing on a similar sort of soil; all are subjected to the same sort of climate. Together they make up a recognizable community of plants.

According to the scale of the plant community, a number of other terms can be used. A plant formation refers to a community in very broad terms. The great belts of vegetation found throughout the world — the rain forests, desert vegetation, deciduous forests, coniferous forests — are examples. Each plant formation breaks down into a number of subsidiary categories. An oak-wood, for example, is a smaller community falling within the deciduous forest plant formation. For such a community, dominated by a single species (i.e. the oak tree) the term consociation is used. If there are two or more species of

equal importance, then the community is called an association (e.g. mixed oak-ash woods). Finally, very small but distinct communities may occur inside associations or consociations. Within an oakwood for instance, there may be a local predominance of ash trees. This lowest category is referred to as a plant society.

Composite flower A plant whose blooms are heads made up of many tiny flowers called FLORETS. There are about 20,000 species, including daisies, dandelions, sunflowers and thistles.

Compound eye, *see* EYE

Compound leaf One that is divided into separate leaflets.

Conch Various marine molluscs with a spiral one-piece shell ranging in colour from white to red. The queen conch found in tropical waters has a shell 30 cm long.

Condor Condors are large American 'vultures'. They are not related to the true vultures of the Old World, although they look very much like them. They have naked heads and necks and wicked-looking hooked beaks. Like the true vultures, they are carrion feeders. There are two species, the Californian condor and the Andean condor. The former is very rare, but the Andean condor is found in the Andes for nearly the whole length of South America. Both birds have wingspans of about 3 metres and are among the largest of all flying birds.

Cone (of eye), *see* EYE

Two different types of plant community. The amount of shrub and ground layers depends on the amount of shade cast by the dominant trees.

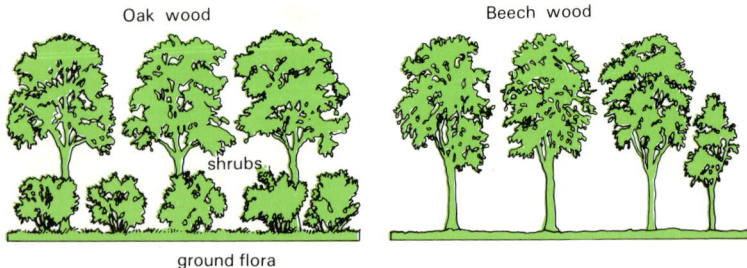

Oak wood

shrubs

ground flora

Beech wood

Conger This is a stout-bodied marine eel, normally about 2.5 metres long. The body is brown or silvery and it is scaleless. The conger lives near the coast in many parts of the world. It is a fierce animal, eating more or less any animal that it can catch. It often hides among the rocks and darts out to catch its prey with powerful jaws and formidable teeth.

Congo eel, *see* AMPHIUMA

Conifer Any tree or plant that bears its seeds in cones. Most conifers have small, evergreen needle-like leaves. Conifers make up an important part of the vegetation in the cooler parts of the world.

Coot

new female cone

male cones

mature female cone

Conifer (pine)

Conjugation A type of reproduction found in some protozoans such as *Paramoecium*, and also in some green algae such as *Spirogyra*. It involves partial or complete fusion of two individual cells and the direct exchange of nuclear material (genes).

Connective tissue The term applied to various body tissues all of which bind cells and organs together. Connective tissue underlies the skin, surrounds nerves and muscles, joins bones and muscles to each other, and often stores fat in its cells.

Contractile vacuole A feature of many protozoans, especially freshwater species. There may be more than one in each organism and they gradually swell as they extract water from the surrounding protoplasm. Suddenly the vacuoles burst and force the water out of the body. They then begin to fill up again. This may be a means of discharging waste from the body but the main function appears to be the removal of water which continually passes into the body because of OSMOSIS.

Cony, *see* HYRAX

Coot Water birds related to the moorhens. They are dark in colour, but can be recognized by the brilliant white beak and the white shield that extends above it. The coot does not have webbed feet, but its toes are provided with flaps which act as paddles in the water.

wild

cultivated

bottle cork

Cork oak

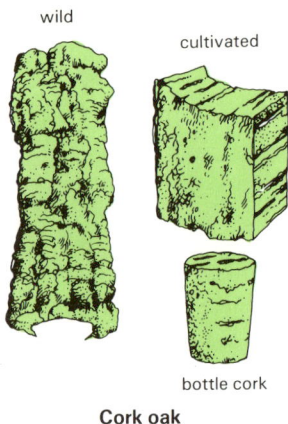

Copepod A group of tiny freshwater and marine crustaceans which form an important part of the plankton. Most copepods have a tapering body with one central eye at the front and two bristly forks at the rear.

Copperhead A small group of venomous North American snakes. Most individuals are less than one metre long and bear live young. Copperheads hibernate during the winter in dens.

Copra, *see* COCONUT PALM

Coral Small animals related to sea anemones and differing from them mainly in possessing a hard chalky skeleton. Some corals live by themselves, but the majority are colonial creatures. Their skeletons are joined together in great masses and they often form coral reefs. All corals live in the sea, and reef-forming kinds live only in warm waters.

Coral snake This name is given to many strikingly coloured snakes, with patterns of rings running around the body. The snakes are found in the Americas and in Africa, Asia and Australia. They are related to the cobras and possess a highly potent venom.

Coriander An annual herb of the Mediterranean area, the coriander is a member of the carrot family. It grows about 90 cm tall, and has much-divided leaves and heads of small white flowers. The dried seeds are used as a spice.

Cork oak An evergreen species of oak growing in Portugal and around the Mediterranean. It has very thick bark, the outer layers of which are periodically removed to make the cork of everyday use.

Corkwing, *see* WRASSE

Corm A short swollen underground stem that carries buds. It acts as a storage organ and also as an agent of vegetative reproduction. The food is stored in the swollen stem and not in leaves as it is in BULBS.

Cormorant A group of large sea-birds, generally with glossy black plumage, although some species have white underparts. The beaks are slightly hooked. The birds live around the coasts and feed almost entirely on fish, which they catch by diving after them. On land, the cormorant often stands with its wings outstretched as if it is hanging them out to dry. The shag is a species of cormorant also known as the green cormorant.

Corn In Britain, a general term for CEREALS or their grain; in North America, the usual name for MAIZE.

Corn-borer, *see* MAIZE MOTH

Cornea, *see* EYE

Corolla The petals of a flower.

food store
used up during
flowering

after flowering
new corm forms
above old one

swollen
stem

new
corm

old
corm

Corm (crocus)

Cormorant

Cortex The region in a stem or root between the epidermis and the vascular or conducting tissues. This name is also given to the outer regions of various animal organs, such as the BRAIN and the ADRENAL GLANDS.

Corymb A type of inflorescence in which there is a head of flowers all at the same level but with different stalk lengths.

Costal Concerning the ribs. The term also refers to the front border of an insect wing.

Cotinga A group of often brilliantly coloured birds living mainly in the rain forests of South and Central America. One of the most spectacular is the COCK-OF-THE-ROCK.

Cotton A bushy plant, found throughout the warmer parts of the world, which grows from 90 cm to nearly 2 metres in height. It has large lobed leaves and flowers ranging from red to cream. The boll, or fruit, contains long white fibres used to make fabric.

Cotton grass Any of 15 species of grass-like SEDGES, found in the northern hemisphere. They grow on boggy ground and produce fluffy white fruiting heads.

Cottonmouth water moccasin, *see* MOCCASIN

Cottontail A small rabbit ranging from southern Canada to South America. The ears are short and rounded and the tail is white underneath. The young are born in shallow depressions in the ground, rather like those of the hare. Young cottontails are blind and helpless at first.

Cotyledon A seed leaf. Cotyledons may or may not emerge from the seed as it germinates. Flowering plants have either one (monocotyledons) or two cotyledons (dicotyledons): conifers may have several.

Couch grass Also known as twitch or spear grass, couch grass is a persistent weed in cultivated ground. It is a species of GRASS which spreads rapidly by underground runners.

Cougar, *see* PUMA

Courgette, *see* MARROW

Courser A long-legged, plover-like shore-bird, living in the Old World from Africa to Australia.

Cowbird A small American bird, of which there are several species, with dark glossy plumage and a finch-like bill. Cowbirds get their name from the way they follow cattle to feed.

Cowrie A carnivorous mollusc related to the sea snails but with the shell whorls enveloped during growth until only the final whorl is visible. Most cowries live in the Indian and Pacific oceans.

Cowslip A perennial herb of Europe and western Asia, related to the PRIMROSE. It has wrinkled, spoon-shaped leaves growing at ground level, and clusters of drooping deep yellow flowers. It grows in meadows.

Coyote A relative of the wolf, the coyote is slightly smaller, measuring about one metre from nose to tail. The fur is tawny coloured and it has a bushy black-tipped tail. Coyotes used to live all over the western part of North America but their range has decreased considerably over the last century. They eat a variety of food including rabbits, small rodents, insects and vegetable matter.

Coypu A large rodent related to the porcupines. Its native home is among the swamps of South America, but it has been reared on farms for its fur and is now established in the wild in North America and Europe. The coypu is well adapted for life in the water with

65

Coypu

webbed hind feet and a soft waterproof underfur — sold as 'nutria' by the furriers. It eats the aquatic and waterside vegetation.

Crab A crustacean with a broad flat exoskeleton or carapace, a short abdomen bent under the body and five pairs of legs, the first pair being larger than the rest and ending in nippers or claws. Crabs have eyes on movable stalks which can be drawn in beneath the carapace. Most crabs are marine but some live in fresh water and a few spend most of the time on land. Many crabs are caught for food in various parts of the world but the name edible crab is normally used for the heavily-built reddish-orange crab of European shores.

Crab apple The wild ancestor of the cultivated APPLE. It is a tree growing up to 9 metres tall, producing small, tart fruit.

Crabeater seal One of the most common antarctic seals reaching some 3 metres in length. Despite its name, the crabeater seal feeds mainly on krill — the minute crustaceans which abound in the surface waters.

Crab-eating fox A South American member of the dog family which is fox-like both in appearance and behaviour. As the name implies it does eat freshwater crabs, but small mammals, frogs and insects are more typical food items.

Crab-eating monkey A common macaque monkey of South-East Asia measuring 50 cm in the body with a slightly longer tail. In swampy coastal areas this monkey supplements its diet of fruit and other plant food with crabs and bivalve molluscs. It also lives in mountain areas, in the jungle and along rivers.

Crab plover A black and white wading bird found along the shores of the Indian Ocean. It feeds mainly on crabs and shellfish which are smashed open against a rock.

Crab spider So called because of the curvature of their legs and the way they scuttle sideways, crab spiders are found almost everywhere. They do not make webs but lie in wait for their prey and then pounce on it. They are sometimes found in flowers, the colours of which they often match perfectly.

Crake These birds are members of the rail family with dull coloured plumage and a short stubby beak. They are found in many parts of the world and are skulking birds living in thick cover and flying only short distances except when on migration. Many species are swamp-dwellers, feeding on water snails, insects and plants.

Crab spider

Cranberry A low-growing creeping evergreen shrub of Europe and North America. It has small leaves, pink flowers and small red acid fruits. It grows on damp moorlands.

Crane A group of rare, long-legged birds, the largest of which stands about 2.5 metres high. They generally breed in lonely marshes and most species migrate long distances from summer

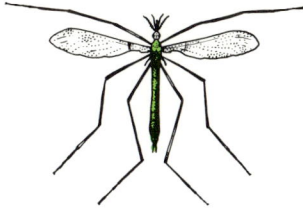

Crane-fly

to winter quarters. They are found in all continents except South America. Cranes feed on leaves, fruit and insects. They have very loud voices which carry over a kilometre.

Crane-fly Popularly known as daddy-long-legs, crane-flies resemble large mosquitoes. Their larvae live in the soil, some of them being the leather jackets which damage crops.

Cranial Concerning the head and skull.

Cranium The skull, especially that part surrounding the brain.

Crayfish The crayfish is a freshwater crustacean, rather like a small lobster. There are several species distributed widely over the world, although absent from Africa. The largest species lives in Tasmania and may weigh up to 4 kg. Crayfishes eat a variety of small water creatures, and some actually leave the water at night to feed on nearby vegetation.

Creeper The name given to various sparrow-sized woodland birds which climb tree-trunks, probing the bark with a finely-pointed beak in search of grubs and insects. The common tree-creeper, known in North America as the brown-creeper, is found in deciduous woodlands throughout much of the northern hemisphere.

Creeping buttercup, *see* BUTTERCUP

Cress A popular salad vegetable, originally native to Iran. It is usually eaten as a seedling. The leaves have a pungent flavour.

Crested newt, *see* NEWT

Cricket A family of insects related to grasshoppers. They have long hind-legs, used for jumping, and long slender antennae. The males produce a shrill 'song' by rubbing the bases of their wings together. Bush crickets, often wrongly called long-horned grasshoppers, are quite similar to the true crickets, although their bodies are less flattened. Crickets eat more or less anything and are often common on rubbish dumps. Most live in warm places.

Croaker The name given to a widely distributed family of fishes remarkable for the noise they make. This ranges from humming and purring to hissing and croaking. The sounds are produced by vibrations of the muscles surrounding the swim-bladder. The Atlantic croaker and the meagre or weakfish of warm waters are two common members of this family.

Crocodile A group of large scaly reptiles inhabiting rivers and estuaries in the warmer parts of the world. Some specimens exceed 6 metres in length. They do not usually go far from the water and spend a lot of time basking in the sun. Young crocodiles feed mainly on insects, but they turn to eating fishes as they get larger. Adults more often catch birds and mammals that come to drink at the water's edge.

Crocus A perennial plant of Europe and Asia, growing from a CORM. It has grass-like leaves, and its yellow, white or purple cup-shaped flowers spring straight from the ground. Some species flower in spring and others in autumn.

Crop Part of the alimentary canal where food is stored prior to completion of digestion. Not all animals have a crop, but it is well developed in grain-feeding birds such as the pigeon.

Crossbill Parrot-like finches in which the tips of the beak are crossed. Females are olive green but males are bright red, marked with black. They

Crossbill

are found in coniferous forests over most of the northern hemisphere and feed almost entirely on conifer seeds.

Cross-fertilization The joining of male and female gametes from different flowers of the same species.

Cross-pollination The transfer of pollen from the stamens of one flower to the stigma of another of the same species.

Crow The crow family contains about 100 species of bird, including the raven, rook, magpie and jay. Of the crows themselves, the carrion crow is the best known. It is about 50 cm long and appears black all over, although close inspection reveals blue and purple in the plumage. It eats almost anything, from grass and fruit to small mammals.

Crowfoot A European and North American aquatic plant of the BUTTER-CUP family. Most of the plant is under water. Its white, buttercup-sized blooms grow above the surface. There are many species.

Crowned eagle Named for its black and white crest, the crowned eagle is found over much of southern Africa, particularly in forested areas. It is a powerful bird whose prey includes monkeys and even small antelopes.

Crustacean An arthropod belonging to the class Crustacea which includes LOBSTERS, CRAYFISH, CRABS, SHRIMPS, BARNACLES, and countless smaller creatures such as COPEPODS and DAPHNIA. Most crustaceans live in water and particularly in the sea, but a few, such as the WOODLOUSE, live on land in damp places. Crustaceans have an exoskeleton which may be fine and transparent as in Daphnia or thick and tough as in the 'shell' of a crab. Most crustaceans breathe through gills.

Cuckoo The European cuckoo is a greyish hawk-like bird which migrates thousands of kilometres between Europe and tropical Africa. It is a social parasite, laying its eggs in the nests of other birds. There are many other birds in the cuckoo family, but not all of them lay in other birds' nests, and only the male European cuckoo has the characteristic 'cuc-coo' call, which has given the name to the whole family.

Cuckoo-pint, *see* ARUM

Cuckoo-shrike Bearing little resemblance to cuckoos or shrikes and related to neither, this group of insect-eating birds is distributed through tropical Africa, southern Asia and Australia. One of the best known is the black-faced cuckoo-shrike of Australia, locally called the bluejay or shufflewing.

Cuckoo wrasse, *see* WRASSE

Cucumber A trailing or climbing hairy-stemmed plant, originally from southern Asia. It is cultivated for its long, pulpy green fruits. It has triangular leaves and yellow or white flow-

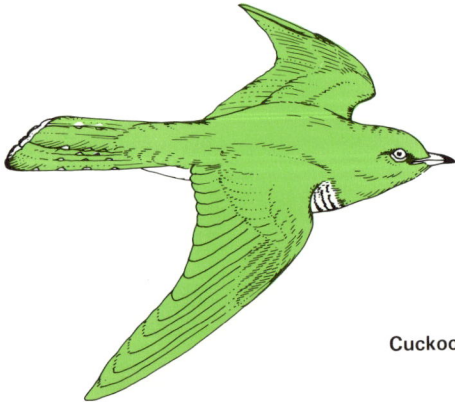

Cuckoo

ers, each one either male or female. Young fruits are often called gherkins, although the true gherkin is the fruit of a related plant from the West Indies.

Cup fungus An ascomycete fungus in which the fruiting body is a shallow cup-shaped disc, often brightly coloured. The majority of cup fungi live on the ground, though some grow on dead wood.

Curassow A small group of birds ranging in size from a pheasant to a turkey. Curassows are readily identified by their horny 'helmets'. They live in Central and South America and spend most of their time in the trees. They feed mainly on fruits and leaves.

Curlew

Curlew The common curlew is a large wading bird, with streaky brown plumage and a down-curved 12-cm beak. It breeds across Europe and Asia, frequenting moorlands and marshes. The curved beak is used for digging small animals from the soil or mud. Several other species breed in the northern hemisphere, and some cover huge distances in their annual migrations to the southern hemisphere.

Currant A low flowering shrub, found wild in the northern hemisphere and parts of South America. It has small round juicy berries. Many cultivated varieties are grown for their black, red or white berries; others are grown as ornamental shrubs. The dried 'currants' used in cake-making are actually small dried grapes.

Cuscus An Australasian tree-dwelling marsupial about one metre long including its long prehensile tail. The various species are all nocturnal and they feed on fruit, leaves and insects.

Cuticle A protective, non-cellular outer layer of a plant or invertebrate animal produced by the epidermis. A typical example is the cuticle of an arthropod which is mainly made up of the protein chitin.

Cutting, *see* PROPAGATION

Cuttle-bone, *see* CUTTLEFISH

Cuttlefish A marine mollusc resembling a squid. Cuttlefishes vary in length from about 4 cm to 1.5 metres. The head bears eight arms and two long tentacles, all with horny suckers. Inside the body there is a chalky 'shell', often washed up on the beach as the 'cuttle-bone'. Cuttlefishes feed on small fishes, which are caught by the tentacles. The animals are famed for their ability to change colour.

Cycad Any of a group of primitive tropical plants, more common in prehistoric times than now. They have thick, woody stems, large fern-like leaves, and bear seeds in cones.

Cyclamen Belonging to the primrose family, cyclamen is a perennial flowering-plant of Europe and south-western Asia. It grows from a corm and bears five-petalled flowers on stems up to 30 cm tall. The petals are strongly reflexed and generally pink. There are several cultivated species.

Cyme A type of INFLORESCENCE where each stem ends in a flower, after giving off one or more side branches.

Cypress Any of 13 species of evergreen coniferous trees of the north temperate zone. Their height ranges from 1.2 metres to 24 metres. Their leaves are like overlapping scales and their cones are rounded.

Cytology The study of cells.

Cytoplasm The fluid contents of cells other than that contained in the NUCLEUS.

D

Dab A small flatfish of shallow coastal waters. It reaches a length of about 40 cm and is brown in colour. It can be recognized by the sharp bend in the lateral line just behind the gill cover.

Daddy-long-legs This name is given to several long-legged CRANE-FLIES common in gardens and grasslands. The larvae are known as leather-jackets, and they feed on plant roots. Some of the long-legged HARVESTMEN (relatives of the spiders) are also called daddy-long-legs.

Daffodil A hardy spring flower with large trumpet-like yellow blooms. It grows from a bulb and is a type of NARCISSUS. The leaves are long and slender. There are many varieties.

Dahlia A member of the daisy family, the dahlia is a tuberous-rooted garden plant, evolved from a Mexican wild species. It produces large, showy composite flowers in every colour except blue.

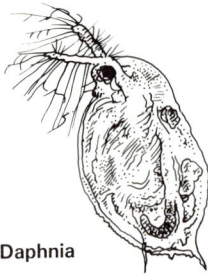

Daphnia

Daisy The name given to many flowers of the family Compositae with petal-like rays. The common daisy frequently seen in lawns is a biennial with leaves growing close to the ground and white, pink or red flower-heads borne on short stalks. The name was originally 'day's eye', because the daisy closes its petals at night or on dull days.

Damselfish A group of small brightly-coloured fishes found mainly in tropical waters, particularly along coral reefs. One of the best known is the clown fish which shelters among the stinging tentacles of a giant sea anemone. Each fish becomes gradually acquainted with, and accepted by, a particular anemone; if the fish enters the tentacles of a different anemone it is liable to be stung to death.

Damselfly A slender-bodied relative of the dragonflies. The front and hind-wings are both the same size and shape. The body is often brightly marked with blue. Young damselflies, like young dragonflies, live in water.

Damson A small variety of PLUM, with usually purple, slightly acid fruit.

Dandelion A low-growing perennial herb of all temperate regions. It has long, toothed leaves, eaten in salads, and yellow composite flower-heads borne on long hollow stalks. Each floret produces a fruit with a feathery 'parachute'. The dandelion root can be used as a substitute for coffee and in some areas varieties of dandelion are used to provide a latex rubber substitute.

Daphnia Also known as the water flea, this is a small transparent crustacean, the largest species being less than 1 cm across. Daphnia appear as dancing specks near the surface of the water and may even form a scum on the surface of ponds. They form the staple diet of many fishes.

Darter Also known as snake-birds, darters resemble cormorants but have longer necks often carried in an S-shaped manner. They inhabit freshwater lakes and rivers where they hunt for fish. Swimming underwater, the birds impale their prey with a rapid jab of their spear-like bills. Darters are found in the warm temperate areas of the world and in America they are known as anhingas.

Darwinism The theory of evolution based on the idea of NATURAL SELECTION, first put forward by Charles Darwin and, independently, by Alfred Russel Wallace.

Darwin's finches A group of birds living in the Galapagos Islands that helped Charles Darwin to arrive at his

Darter

theory of evolution by natural selection. Several species live in the various islands, all derived from a few finches that were blown to the Galapagos from South America. Before they arrived there were few birds on the islands, so the newcomers were able to evolve into many different forms, each having a different way of life.

Dasyure The dasyure is the 'native cat' of Australia. It is not a true cat, but a pouched mammal with short legs and a long bushy tail. Dasyures range from about 30 cm to over one metre in length.

Date palm A kind of palm with a tall stem and a feathery crown. It bears large bunches of rich red fruits which are often dried and eaten. The tree originated in North Africa, and grows to more than 30 metres in height.

Dead-nettle A group of herbs which superficially resemble stinging NETTLES, with square stems and rough leaves. But they do not sting, and have red, white or sometimes yellow flowers in whorls at the bases of the upper leaves.

Death cap, see AGARIC

Deciduous (of plants) Shedding leaves seasonally. (of insects) Shedding wings after mating. (of teeth) Milk teeth.

Decurrent (of a leaf) With the base extended and running down the stem as a 'wing'. The gills of some toadstools are also decurrent.

Decussate Arrangement of leaves on a stem in which the leaves are in opposite pairs and in which each pair is at right angles to the pairs above and below.

Deer The general name given to ruminants bearing antlers rather than horns. In all species, with the exception of reindeer, only the male bears antlers. The antlers are bony, usually branched, and renewed each year.

Deer mouse American rodents very similar to the European fieldmouse, although the two are not closely related. There are more than 50 species ranging from 12–40 cm in length including the tail. The feet are white, and the animals are often called white-footed mice.

Dehiscent Fruits that split open and release their seeds are said to be dehiscent. The poppy capsule is a good example and so is the pod of the pea.

Dentine, see TOOTH

Deodar A kind of CEDAR which grows in the western Himalayas. It has a distinctly pyramidal shape. The deodar can be distinguished from the other cedars by the distinctly drooping leading shoots and the shoots at the tips of its branches.

Dermis The lower layer of SKIN.

Desman An aquatic, mole-like, insect-eating mammal with a long, mobile snout. There are two species, one living in Russia and the other in northern Spain.

Devil fish Related to skates and rays, these fishes live in the warmer seas of the world. The front fins are greatly enlarged as 'wings' and they also have forward-pointing extensions at the front of the head. These look like horns and give the fishes their common name. The largest of the species is the manta ray, which may be 8 metres across and weigh nearly 2 tonnes.

Devil's coach-horse, *see* ROVE BEETLE

Dhole This is the wild dog of India. It is not a true dog, and can be distinguished by its rounded ears and short muzzle. The fur is greyish-brown and the bushy tail is tipped with black. Dholes range from India and Malaysia to China and parts of the Soviet Union. They hunt in packs of 100 or more individuals, running down deer and other animals.

Diadem spider, *see* ORB SPIDER

Diamondback rattlesnake, *see* PIT-VIPER

Diana monkey, *see* GUENON

Diaphragm The sheet of muscle separating the thorax from the abdomen in mammals.

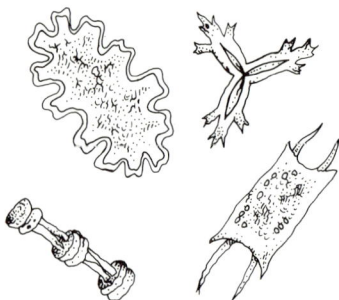

A variety of diatoms

Diatom A group of algae consisting of unicellular or colonial forms abundant in fresh water and in the sea.

Dibatag A rare, gazelle-like antelope whose range is confined to tall grasslands in parts of Somalia and Ethiopia.

Dichotomous Dividing regularly into two equal branches.

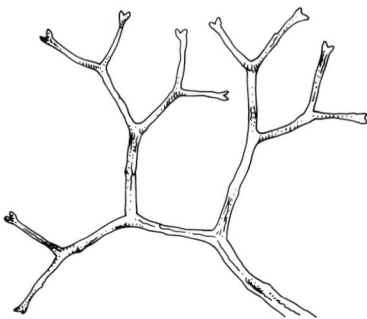

Dichotomous branching

Dicotyledon A kind of flowering plant which possesses two seed-leaves, known as cotyledons. Most flowering plants are dicotyledons. Their mature leaves are generally fairly broad and have a distinct network of veins, and their flowers generally have four or five petals (see Monocotyledon).

Differentiation The change in the structure of cells and organs during their development as they become specialized to perform a particular function.

Digestion The breakdown of complicated food substances into simpler compounds which can be used by the animal to build up its own body and to provide energy. Digestion is performed by enzymes, mainly in the food canal, and the products are absorbed into the body.

Digitalis, *see* FOXGLOVE

Dik-dik A small antelope distributed over most of Africa. The males have short horns, partly hidden in a tuft of hairs on the head. The tail is a mere stump, and the animal also has a longish snout.

Dill A small European herb related to parsley. It has bitter seeds and tiny leaves, which are both used as flavourings.

Dimorphism The existence of a species in two distinct forms. Sexual dimorphism, in which males and females differ, occurs to some extent in all bi-sexual animals of course, but it is particularly marked in many birds and insects. Some insects also show seasonal dimorphism, with spring broods quite different from summer ones.

Dingo This is the wild dog of Australia — one of the few placental (non-pouched) mammals to arrive before the Europeans. It arrived with the Aborigines when they reached Australia from southern Asia. It is a reddish-brown animal, about 50 cm high at the shoulder. The ears are erect and pointed, and the tail is bushy. Dingoes live in small packs and hunt sheep, wallabies and kangaroos.

Diploid Having a double set of CHROMOSOMES in the cell nucleus. Normal body cells are diploid. Sex cells are formed by a special process in which only one set of chromosomes goes to each cell. Cells with only one set of chromosomes are called haploid. When sex cells join together at fertilization the diploid number is regained in the resulting zygote or embryo.

Dipper A group of wren-like birds with short wings and tails. They live by the sides of swift-flowing, boulder-strewn streams and feed mainly by catching insect larvae under the water. Dippers mostly live in the northern hemisphere.

Distal The far end, away from the body. For example, the hand is at the distal end of the arm.

Diver The divers, or loons, are streamlined water birds with very short tails and straight pointed beaks. They are

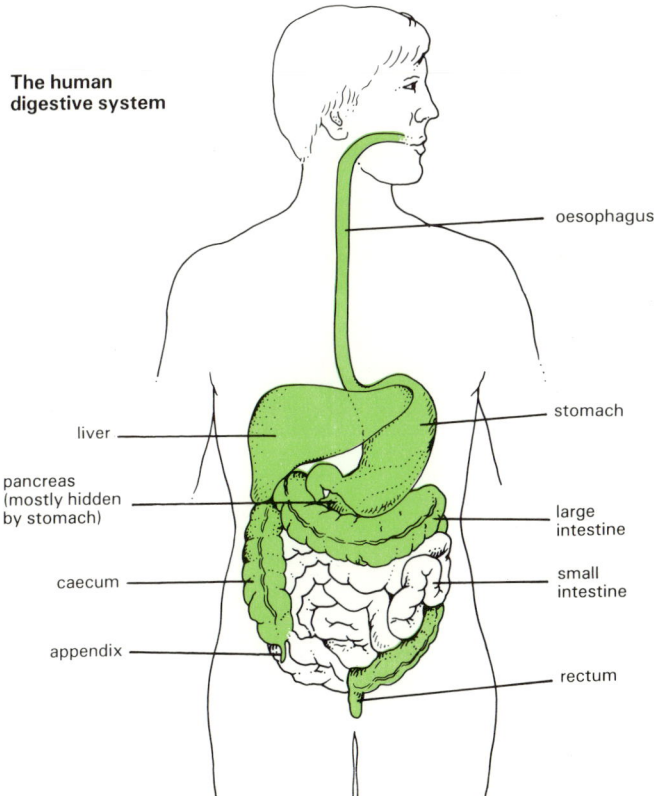

The human digestive system

oesophagus

stomach

liver

pancreas (mostly hidden by stomach)

large intestine

caecum

small intestine

appendix

rectum

excellent swimmers but can only shuffle on land. Divers live in the northern parts of the world. They spend the winters in coastal waters, but move to inland waters to breed. They feed mainly on fishes.

Diving beetle The popular name for a number of carnivorous aquatic beetles which spend much of their time under water, particularly the great diving beetle and the whirligig beetles.

DNA (Deoxyribose nucleic acid) A substance found in the chromosomes of cell nuclei. DNA carries the GENES which are responsible for passing hereditary 'instructions' from one cell generation to the next.

Dock Any of a group of plants with stout tapering roots, large flat leaves and greenish flowers which often turn red as they develop into fruits. There is a papery sheath, called the ochrea, at the base of each leaf. Docks are common weeds. Despite their bitter taste, some are eaten as vegetables.

Dodder Any of a group of leafless parasitic plants which attach themselves to other plants by suckers. Dodders have thread-like red or yellow stems — and produce clusters of tiny funnel-shaped flowers.

Dog, domestic Dogs have been domesticated for at least 8000 years. They have probably been derived from mainly wolf stock with a possible jackal admixture. Domestic dogs of all shapes and sizes have been bred by man over the centuries, ranging from the St Bernard, weighing as much as 90 kg, to toy dogs of less than 1 kg in weight.

Dogfish This name is given to several species of small sharks and cat-sharks. One of the commonest is the lesser spotted dogfish. This fish is about one metre long, including the slender tail, and it has extremely rough skin. The dogfish is an important food fish. When fried, it is known as rock salmon or rock eel.

Dogwood Any of about 40 shrubby trees of Eurasia and North America. They have small oval leaves and creamy flowers, and produce small black berries.

Dolphin A group of small toothed whales with a pronounced beak. Best known are the common dolphin and the BOTTLE-NOSED DOLPHIN. The common dolphin grows to about 2.5 metres in length. It is black above and white underneath. Common dolphins roam the oceans in schools, travelling at speeds up to 40 km/h.

Dolphin fish The name given to two fast-swimming tropical fishes which feed mainly on flying fish. They can reach speeds of 60 km/h or more.

Dominant (1) A GENE that overrules the action of another affecting the same feature(s). (2) The major plant in a community.

Donkey A domesticated ass long used in poorer parts of the world as a beast of burden and for pulling carts and ploughs.

Dormant In a resting condition — e.g. hibernating.

Dormouse A rodent widely distributed over Europe and Asia. There are several species. The largest is the edible dormouse or glis-glis, a grey squirrel-like animal about 30 cm long. Most dormice live among the woodland undergrowth, although the glis-glis prefers taller trees. The animals feed mainly on fruits and nuts. They hibernate during the winter.

Dorsal On or concerning the side of an animal that is normally uppermost.

Dotterel A small wader inhabiting the Arctic tundra and the mountains of Europe and Asia. The dotterel's white and brown plumage merges well with the ground in the bleak areas where it lives.

Double coconut, *see* COCO DE MER

Douglas fir One of the tallest coniferous trees, more closely related to the SPRUCES than the true FIRS. It grows up to 60 metres tall, and has soft flat needles with white lines underneath and a sweet smell. It has ragged egg-shaped cones which do not break up when they are ripe.

Dolphins are natural acrobats and have a great sense of fun. They are also very friendly towards humans and have even been known to help save the lives of drowning swimmers. They are not so friendly towards sharks though and will often attack and kill them.

Douroucouli This is the world's only nocturnal monkey. It is also called the owl monkey because of its large eyes and rather owl-like face. Douroucoulis live in the rain forests of South America and feed mainly on small animals.

Dove The name given to any species of bird belonging to the family Columbidae. Some are called pigeons, though there is no real distinction between doves and pigeons.

Dragonet A small bottom-living fish whose name is derived from the enlarged dragon-like fins and spines of the brightly-coloured male. The female is inconspicuous by comparison and was once thought to be a separate species.

Dragonfly

Dragonfly A group of large colourful insects with huge compound eyes and a powerful pair of jaws. Using their keen eyesight and rapid flight, dragonflies catch other insects on the wing. Their water-living nymphs prey on other animals, including tadpoles and even small fishes.

Driver ant, *see* ARMY ANT

Dromedary, *see* CAMEL

Drongo A group of insect-eating birds living in the tropical forests of Africa and Asia. Most have conspicuous bristles around the nostrils and long, ornately-shaped tail feathers.

Drupe A fleshy fruit, the inner layer of which is hard and woody and normally encloses a single seed (e.g. plum). The outer layer is occasionally fibrous or leathery instead of fleshy, as in the coconut and the walnut.

Duck A waterfowl belonging to the family Anatidae which also includes the geese and the swans. Ducks fall into two groups: the freshwater surface-feeding ducks (such as the MALLARD, MANDARIN, PINTAIL, SHOVELER and TEAL) and the sea or diving ducks (such as the EIDER, MERGANSER, POCHARD and STEAMER DUCKS).

Duckweed The name given to a group of very tiny flowering aquatic plants. Duckweed floats on the surface of ponds, and consists of a flat disc or frond with a long thin root. The flowers are almost invisible. Reproduction occurs mainly by producing detachable buds which grow directly into new discs.

Dugong A seal-like mammal about 2.5 metres long, with flipper-like front limbs and a broad paddle for a tail. Dugongs live along swampy tropical coasts of the Indian Ocean, from Africa to northern Australia. They feed on eel grass and other aquatic vegetation. Dugongs are closely related to the manatees and both groups are often known as sea cows.

Duiker A small, short-legged antelope with an arched back. The tallest of the dozen or so species is only about 60 cm high at the shoulder. Both sexes have short, pointed horns. Duikers are found throughout most of Africa.

Dung beetle, *see* SCARAB

Dunnock, *see* ACCENTOR

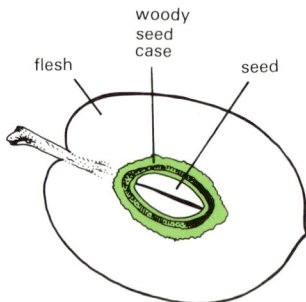

Drupe (plum)

E

Eagle A large broad-winged bird of prey. Examples include the BALD EAGLE, the CROWNED EAGLE, the FISH EAGLE, the GOLDEN EAGLE, the HARPY EAGLE and the MARTIAL EAGLE.

Ear The organ which, in vertebrates, is concerned with the detection of sound and also with the sense of balance. 'Ears' also occur in other animals such as insects, but these are far less complicated structures and are concerned merely with the detection of sound. The human ear is typical of mammals and differs only in small details from that of most other vertebrates. There are three main regions: the outer ear consisting of the ear lobe or pinna and the passage leading to the ear-drum (tympanic membrane); the middle ear which contains tiny bones transmitting sounds from the drum to the inner ear and which connects with the throat by means of the eustachian tube; and the inner ear which, encased in the bone of the skull, contains the actual sense organs of hearing and balance.

Eared seal, *see* SEAL

Earshell, *see* ABALONE

Earthworm Various segmented worms which burrow through the soil, eating earth and extracting nourishment from it. The excreted earth is often left on the surface as worm casts. Earthworms help to aerate and drain soil. The common earthworm of temperate regions is about 25 cm long; the largest, an Australian species, grows to 3 metres.

The human ear

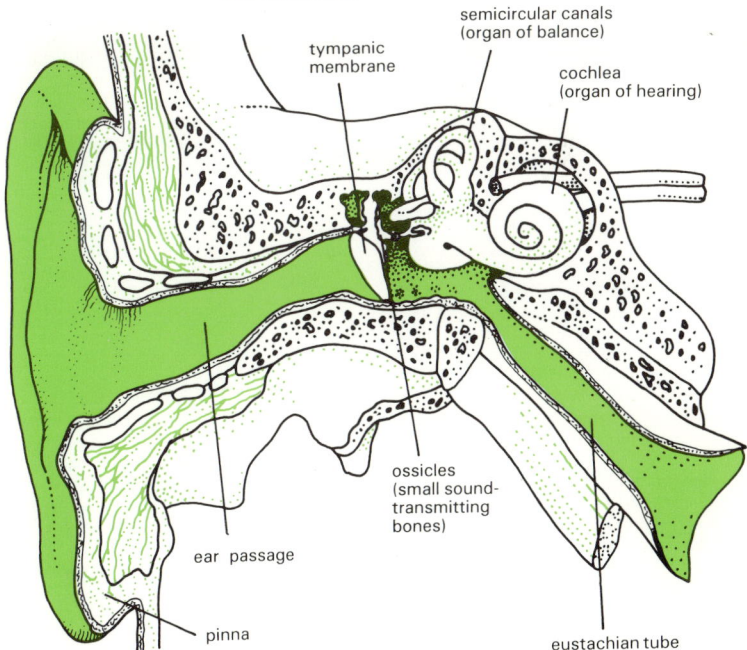

semicircular canals (organ of balance)

tympanic membrane

cochlea (organ of hearing)

ossicles (small sound-transmitting bones)

ear passage

pinna

eustachian tube

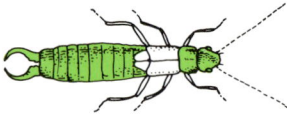

Earwig

Earwig Common scavenging insects easily recognized by their hind pincers, curved in the male, straight in the female. Earwigs hide in crevices during the day and emerge at night. They are rarely seen in flight, though many can fly quite well. As with beetles, their wings are folded away behind hard wing cases when at rest, but many species are wingless.

Ebony The name of a group of trees related to PERSIMMONS, which are valued for their hard, black wood. The trees which yield this timber grow in India, Sri Lanka and Angola. They are up to 25 metres tall.

Echidna, *see* SPINY ANTEATER

Echinoderm A member of the large phylum Echinodermata. This consists entirely of marine animals including the starfishes and sea urchins. They are radially symmetrical and the outer covering contains numerous chalky plates, sometimes produced into spines, which give the animals a rough appearance. There is no brain, nor is there even any structure that can be regarded as a head. The nervous system is merely a network of nerves with thicker strands in places. A feature unique to these animals is a system of water-filled canals which runs in the body. Tiny branches reach the surface and are known as tube feet. They are used for moving about and also aid respiration. In general, the sexes are separate but a few species are hermaphrodite. The sex-cells are usually shed freely into the water where fertilization occurs. The young stages show similarities with certain chordates, rather than with any other invertebrates.

Ecology The study of animal and plant communities and the ways in which they react with each other and to any changes in the environment.

Edelweiss A low-growing herb of the Alps and Himalayas. It has long narrow leaves and composite white flowers. The leaves are thickly covered with white, woolly hairs.

Edentate A group of South American mammals belonging to the order Edentata. They include the armadilloes, ant-eaters and sloths. The teeth are reduced to small pegs in the armadilloes, absent in the ant-eaters and modified for grinding in the herbivorous sloths. The claws are well-developed for digging, or hanging in the case of sloths.

Eel A group of elongated fishes. Most species are naked and slimy, although some have small scales imbedded in the skin. There are both marine and freshwater species. The latter make long migrations to the sea during the breeding season. European and eastern North American eels migrate to an

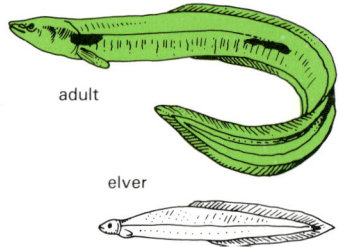

adult

elver

Eel

area near the Sargasso Sea to spawn and the young animals (elvers) may take as long as three years to swim back to the rivers. Eels are carnivorous creatures ranging from 30 cm to 2 metres in length. One of the best known marine species is the moray eel, a fierce fish armed with formidable teeth.

Eelgrass An aquatic plant with long, grass-like leaves. It is among the very few marine flowering plants, and grows in muddy estuaries, providing food for many marine animals.

Eelworm The name given to many small NEMATODES which live in the soil, water and even vinegar. Some are agricultural pests, attacking root crops such as potatoes and sugar beet.

Effector An organ that acts in response to a signal. Muscles and glands are the main effectors. RECEPTORS receive stimuli from outside the body and send messages to the brain. The brain then sends the appropriate message to the effectors which take the necessary action.

Egg-eating snake Many snakes eat eggs, but the six species of egg-eating snakes found in Africa and India live almost exclusively on them. They glide through the tree-tops at night, seeking out birds' nests by smell. The eggs are swallowed whole and are then cut open by a row of teeth, or pegs, in the roof of the throat. The contents of the eggs are swallowed, but the shells are compacted and ejected through the mouth. Egg-eating snakes have no venom.

Egg-plant, *see* AUBERGINE

Egret A group of mainly white birds related to the herons. Most live around lakes and marshes and feed in shallow water on frogs, fishes and other aquatic animals. The cattle egret, however, feeds on land. It forages in small flocks and eats the insects it disturbs in the vegetation. It often associates with cattle and other larger animals.

Eider Famed for its soft warm down, the eider is a duck about the same length as a mallard but considerably stockier. There are four species, all living around the northern coasts of the Pacific and Atlantic. The male common eider is a striking black and white bird but the female is brown and speckled. The birds usually move south for the winter and tend to stay further out to sea than during the summer. They dive for shellfish and crabs on the seabed.

Eland This is the largest antelope, standing up to 2 metres at the shoulder. It is a heavy, ox-like creature and both sexes carry spiral horns up to one metre in length. There is a short mane and a tuft of long dark hair on the throat. The animals live in herds of up to 100 and roam the open plains of central and southern Africa.

Elasmobranch A member of a large group of marine animals (Elasmobranchii) which includes sharks, rays and other fishes whose skeletons are made entirely of cartilage with no trace of true bone.

Elder A small tree or shrub of the northern hemisphere. The bark is thick and soft; the tree has broad compound leaves, and clusters of small, strongly scented white flowers. They produce small black berries. Young shoots contain soft pith. Both the leaves and twigs emit a pungent smell when broken.

Electric fish A number of fishes use electrical impulses to find their way about and some have the ability to generate electric shocks capable of stunning their prey. Among the best

Egret

known are the electric catfish of African lakes and rivers, the electric ray of warm Atlantic waters and the electric eel of South American rivers. The latter, a fish quite unrelated to the true eels, is the most spectacular. Growing to 2 metres or more in length, it can generate a shock powerful enough to stun a horse.

Electric ray, *see* RAY

Elephant The two species of elephant are the largest living land mammals. The African elephant is the larger of the two, reaching 3.5 metres in height and 6 tonnes in weight. The trunk of an elephant is used for carrying food

African elephant

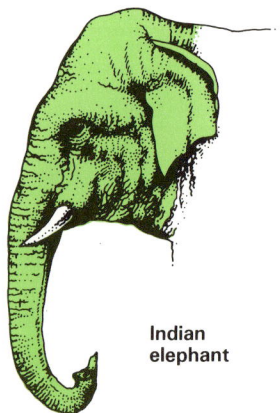

Indian elephant

and water to the mouth, for spraying water over the body, for lifting things, and for smelling. It is immensely strong, and yet is also very sensitive, especially near the tip. The two species differ mainly in the size of their ears: those of the African species are much larger and reach well below the level of the mouth. The African elephant is found over most of the southern half of Africa, while the Indian elephant lives in the forests of south-eastern Asia, from India to Sumatra. They eat a variety of plant material.

Elephant seal Also called the sea elephant, this creature is the largest of the seals. The male can reach a length of 6 metres and a weight of 3 tonnes. Females are much smaller and weigh less than one tonne. The name refers to the peculiar drooping snout of the male. There are two species, both very much alike. The northern elephant seal lives off the coast of Mexico and California, while the southern elephant seal lives mainly in the South Atlantic and the Antarctic. Like most other seals, they feed on fishes and squids.

Elephant shrew Found only in Africa, these insect-eating mammals belong to the same order as the ordinary shrews, although they are not closely related. Elephant shrews get their name from their greatly elongated snouts. They range from 10–30 cm in length, excluding the tail.

Elk In North America this name is normally given to the WAPITI, a large deer. In Europe the name is applied to an animal identical to the North American MOOSE.

Elm The name of a group of large trees. The dense clusters of flowers appear before the leaves and give the trees a red tinge. The leaves are rough and oval, usually asymmetrical at the base. Elms grow up to 30 metres tall and produce tough timber.

Elytron (plural elytra) The hard front wing of a beetle.

Embryo A young animal or plant living exclusively on food provided by its female parent. It may be a young plant in a seed, a chick inside the egg, or a young mammal in the mother's womb.

Embryology The study of embryos and their development.

Emergence, *see* THORN

Emigration, *see* MIGRATION

Emperor moth, *see* ATLAS MOTH

Emperor penguin This is the largest of the penguins, reaching about one

fruit

English elm

metre in height. It is also the most familiar owing to its unmistakable black and white plumage relieved by an orange collar and its waddling gait. Emperor penguins live around the coasts of Antarctica and breed in large rookeries. The female lays one egg which the male balances on his feet for the whole of the two-month incubation period during the cold Antarctic winter.

Emu The second largest living bird, the Australian emu stands about 1.8 metres high. It is related to the

cassowary and shares its drooping downy plumage. The emu cannot fly, but can run at 60 km/h over short distances. It is a vegetarian and is often a nuisance on farms during the dry season.

Enamel The hard white covering of teeth.

Endemic Native to a given area. Especially used of pests and diseases.

Endive A curly-leaved salad vegetable, related to CHICORY, and cultivated since ancient times. The leaves are slightly bitter.

Endocrine gland, *see* GLAND

Endoskeleton, *see* SKELETON

Endosperm A special food-storing tissue surrounding the embryo in the seeds of many plants.

Entomology The study of insects.

Environment All the factors in an organism's surroundings, including soil, other living things, climate, temperature, wind, etc.

Emu

Embryo

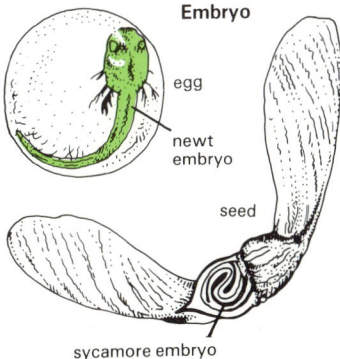
egg

newt embryo

seed

sycamore embryo

Enzyme A complex protein that promotes some reaction in the body. An enzyme is, in fact, an organic catalyst, increasing the rate of a reaction without itself being used up. Without the enzymes, many of the reactions in the body would be so slow as to be unnoticed. All bodily processes — digestion, respiration, etc — rely on enzymes to speed up their rates.

Epidermis The outermost layer of cells (skin) in a plant or animal.

Epigeal Referring to germination in which the seed leaves (COTYLEDONS) come above ground (e.g. cabbage).

Epiglottis A flap of cartilaginous tissue in the throat that closes the wind-pipe opening (glottis) when food is swallowed.

Epigynous Flowers in which the pet-als, etc., arise from above the enclosed carpels. The carpels are inferior (e.g. apple).

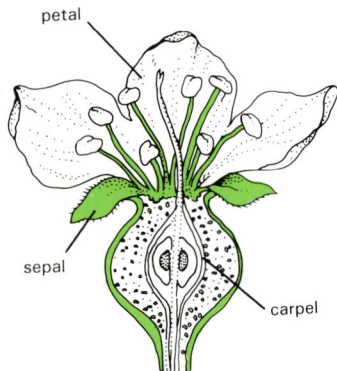

Epigynous flower

Epiphyte A plant growing on another but using it only for support and not drawing food from it (e.g. mosses and orchids on trees).

Epithelium A sheet of cells that lines most cavities of the body and covers the exterior. The skin, the lining of the gut and other organs, such as the lungs and blood vessels, are examples. The cells of epithelia may serve very differ-ent purposes. Those lining the salivary glands and the glands in the intestine for example, produce digestive enzy-mes. The ones forming the outer covering of the skin are mainly protec-tive, while the cells of the lung lining produce the wet mucus in which the oxygen dissolves before passing to the blood. Many epithelia are ciliated — e.g. windpipe — and the cilia help to waft out the foreign particles (dust).

Ergot The common name for the ascomycete fungus *Claviceps* which infests the grains of rye, oats and many other grasses.

Erythrocyte A red blood corpuscle.

Esparto A group of tall wiry grasses, growing in Spain and North Africa. Esparto fibres are used to make ropes, mats and paper.

Eucalyptus The name of about 600 species of evergreen Australian trees, also called gum-trees. Some grow up to 100 metres tall. They have evergreen leathery leaves containing a strongly-scented oil, and hard timber.

Euglena A group of microscopic green plants, often found as scum on the sur-face of still, fresh water. Euglena pro-pels itself through the water by waving a whip-like FLAGELLUM.

Eustachian tube A tube running from the middle ear to the throat of most tetrapods. It enables pressures on each side of the ear drum to remain equal.

Everglade kite A hawk which lives in swamps of South and Central America. It was once common in the Florida Everglades but is now very rare. The everglade kite is believed to feed exclusively on water snails.

Evergreen A plant that bears leaves all the year round, e.g. holly. Leaf fall is not a seasonal process but a continuous one, with just a few leaves being replaced at a time.

Evolution The process whereby living things are believed to have arisen from less advanced forms by gradual change. A great deal of evidence sup-ports the idea of evolution, especially the evidence provided by the fossil record. Darwin and Wallace put for-ward the theory of NATURAL SELEC-TION to explain how these gradual changes could be passed on to succeed-ing generations.

Excretion The removal from the body of waste products — unwanted and often poisonous substances produced by the vast number of biochemical reactions going on in the body. The simple act of breathing out carbon dioxide is a form of excretion, but the term is generally restricted to the removal of nitrogen-containing waste. These result from the breakdown of proteins in the body — proteins from surplus food perhaps, or from old and worn-out tissues that are continually being replaced. All but the simplest animals have special organs to get rid of this waste material. In the vertebrates this kind of excretion is performed by the KIDNEY and associated organs.

Exocrine gland, *see* GLAND

Exoskeleton A skeleton that is outside the body — e.g. crabs, insects and snails. Muscles are attached to the inside.

Eye The organ of sight. There are many different designs, but the eye is basically a lens which focuses light rays from the surroundings onto sensitive cells. When the light rays stimulate the cells, they send signals to the brain, which then 'translates' the signals back into a picture of the surroundings. The simplest eyes may not be able to do more than detect light and dark areas, but human eyes, like those of all the vertebrates, are very complicated and usually very efficient.

The vertebrate eye is essentially a fluid-filled ball with a very tough wall, called the sclera, and a lens suspended near the front. The light-sensitive cells form a layer known as the retina at the back of the eye-ball. The front of the sclera is transparent and known as the cornea, and between this and the lens there is a muscular ring known as the iris. This is the coloured part of the eye. Light rays pass through the cornea and then through the pupil in the middle of the iris. Then they pass through the lens, where they are bent and focused onto the retina. The lens is relatively soft and its shape can be altered so that it can focus light rays from near and far. Muscles around the inside of the sclera produce these changes in the lens without any conscious control. The muscles of the iris also work automatically, enlarging the pupil in dim light and closing it in bright light so that the right amount of light gets through to affect the retina. The pupil is generally circular, but in some animals, such as cats, it is slit-like in daylight.

The cells of the retina are of two main types, known as rods and cones. Each cell is connected to a nerve fibre, and when light rays strike the cells the

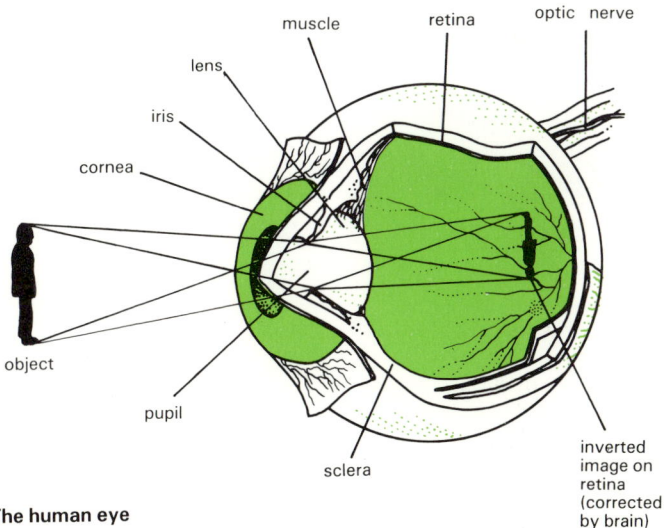

The human eye

fibres carry messages straight to the brain. The fibres are bundled together to form the optic nerve, and the point where this leaves the back of the eye is called the blind spot — because there are no sensitive cells there to receive the light rays. The rods cannot distinguish colour, but they are sensitive to very low light levels and enable a person to see when it is quite dark. The cones are able to distinguish between different colours, but they need plenty of light to work properly. This is why we see only in black and white at night.

Insects and various other arthropods possess compound eyes, built to a completely different design. Each eye has numerous minute lenses called facets, and each lens sends its own image to the brain via its own nerve fibre. The insect thus sees a picture made up of lots of tiny dots — one for each facet receiving light rays. Dragonflies, which chase other insects through the air, need good eyesight and their eyes have up to 30,000 facets, producing a very detailed picture. Many ants, on the other hand, which spend much of their life below ground, have very few facets and some have no eyes at all.

head of insect

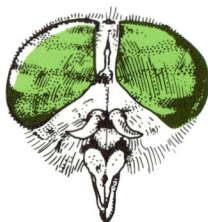

Compound eye

Eyebright A small annual herb of Eurasia and North America found on grasslands as a rule. It has leaves with cut edges, and small white and yellow flowers. Eyebright is a semi-parasite, taking water from the roots of neighbouring plants. There are several species.

F

Faeces Undigested material passed out of the food canal at the anus.

Fairy fly This name is given to various minute wasp-like creatures, not true flies at all. They belong to a group called chalcids and they spend their early lives inside the eggs of other insects. They are parasites. The name fairy fly refers to the tiny wings, little more than a couple of veins and a few hairs.

Fairy shrimp A transparent and almost colourless relative of the true shrimps. It swims on its back by means of rhythmic movements of its many legs, which also function as gills and as food collectors. The fairy shrimp's eggs can survive for many years without water, and this is why the creatures can appear quite suddenly in small temporary puddles.

Fairy tern, *see* NODDY

Falcon The name given to any bird of prey belonging to the family Falconidae. Falcons are generally smaller than other birds of prey, with long pointed wings and long tails. Among others they include the PEREGRINE FALCON, the GYRFALCON and the KESTREL.

Falkland Islands steamer duck, *see* STEAMER DUCK

Fallopian tube One of two funnel-shaped tubes in female mammals leading from the ovaries to the uterus. Eggs are normally fertilized on their way down the tubes.

Fallow deer This attractive deer has been introduced to many parts of the world by man, but it originally came from the Mediterranean region. The coat is reddish-brown, dappled with white spots in summer, and there is a prominent white patch on the rump, extending on both sides of the black tail. The fallow deer is a little under one metre at the shoulder, and the buck carries broad palmate antlers. It feeds mainly on grass.

False fruit A FRUIT formed from parts of the flower other than just the CARPEL (e.g. apple, strawberry).

semi-plume feather

contour feather

Feather

barb

hook

barbule

unconnected barbs (down)

shaft

shaft

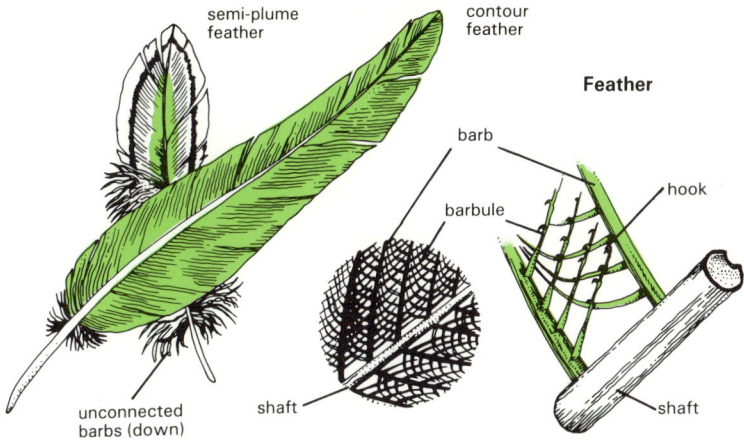

False scorpion A minute animal only distantly related to the true scorpions. Most of the 1500 known species live in leaf litter and other vegetable debris. They feed on small insects and mites which they catch with their relatively huge pink claws. A few species live between the pages of old books or in the bodies of stuffed museum specimens. They feed on the booklice that they find there.

Family A category used in the classification of organisms. Zoological family names always end in -idae; botanical family names generally end in -aceae.

Fanworm This name is given to various marine bristleworms that live in tubes and filter their food from the water with a crown, or fan, of stiff bristles. Some fanworms live in little tubes made of lime, others make tubes of sand grains. One of the best known is the peacock worm, whose sandy tubes often stick out of the mud at low-tide.

Fat-hen A common European weed found in almost all locations ranging from cultivated land to rubbish dumps. It has fleshy leaves, small inconspicuous green flowers and black seeds. It was cultivated as a food crop until the 1800s.

Fauna The animal life of a certain region or time.

Feather An outgrowth of the skin that is found only in birds. Feathers probably evolved from reptile scales and, although they are used for flight now, their first function was almost certainly as a protection against the cold. The large feathers that sheathe the body of a bird as well as covering the tail and wings are called contour feathers. Each consists of a partly hollow, horny central shaft from which soft barbs branch off on either side. The barbs are so closely packed together that they appear to form a continuous surface. In actual fact they are separate units, though each one is attached to its neighbours by a series of hooked and notched barbules. The barbs are not linked in the ostrich and other flightless species. Beneath the contour feathers are down feathers. These are much smaller and softer. Down feathers insulate a bird from the cold by trapping a layer of air against the skin.

Featherstar A relative of the starfish, this creature has five arms which divide into as many as 200 feathery branches. The arms filter food from the water. The small central body is largely covered with plates of limestone. Little claws on the underside of the body anchor featherstars to the seabed, but the animals are able to release themselves and swim away.

Male fern   Adder's tongue   Hart's tongue

Three of the many species of fern

Femur The thigh bone.

Fen A plant community developing in shallow water overlying alkaline rocks. The typical plants are reeds and sedges.

Fence lizard The commonest lizard in the United States is a grey-brown iguana about 25 cm long. Other names for it are the swift lizard and pine lizard.

Fennec This little animal is the smallest of the foxes, yet it has the largest ears. The head and body measure about 40 cm in length and the ears are 15 cm long. The fennec lives in the deserts of North Africa and Arabia. Its pale sandy colour blends well with the desert sands, and its huge ears help it to pick up the slightest sound. The fennec feeds on a variety of small animals.

Fennel A yellow-flowered parsley-like plant of Europe, western Asia and North Africa. Its thread-like leaves and flattened fruits taste of liquorice and are used as flavouring. The tough stem is about 1.2 metres high.

Fer de lance A relative of the rattle-snakes, the fer de lance is found in coastal districts from the Caribbean to Argentina. It is about 1.5 metres long and has a diamond pattern on the back. This is a poisonous snake and many accidents occur because it frequents sugar plantations and houses in its search for rats and mice.

Fermentation The decomposition of organic materials, especially carbohy-drates, by micro-organisms. A good example is the breakdown of sugars into alcohol in the process of wine-making. Yeasts are the organisms concerned here.

Fern A large group of flowerless plants. There are about 10,000 species, from moss-like plants to trees up to 12 metres tall. They mostly have large leaves, called fronds, on which spores form. Ferns reproduce from these spores (see ALTERNATION OF GENERATIONS). They are most abundant in tropical areas, especially in damp habitats.

Ferret A domesticated version of the polecat which has been used for centuries to catch rabbits. The muzzled ferret is sent into a rabbit hole to chase the rabbits out into nets. The blackfooted ferret is a rare North American animal. It is about 30 cm long and, like the ordinary ferret, related to the weasel. It feeds on prairie dogs.

Fertilization The joining together of two sex-cells (gametes) to form a zygote — the first cell of a new generation.

Fescue The name of a group of mostly fine-leaved grasses grown for fodder and as lawns. They produce the springy turf on the sheep-grazed downlands.

Fibula The smaller of the two bones in the lower part of the vertebrate hindleg.

Fiddler crab This is a little crab, only about 2 cm across, found in great numbers on tropical beaches. The males have one very enlarged claw and one normal claw. They are brightly coloured and the large claw is waved about as a call sign to attract a mate. Some fiddler crabs can make a sound rather like that of a cricket by rubbing their claws against a row of teeth on the shell.

Field mushroom, *see* MUSHROOM

Fig A deciduous Mediterranean tree or bush producing pear-shaped edible fruit. Figs have large three- or five-lobed leaves and two kinds of flower, male and female. They are widely cultivated. There are many related plants in other parts of the world including the banyan tree.

Fighting fish Many fishes fight, but the most famous is the Siamese fighting fish. It has been reared for its fighting ability, and fish-fighting is a traditional sport in its native Thailand. The wild fish is yellowish-brown, but many other colour varieties have been produced now.

Fiddler crab (male)

Siamese fighting fish

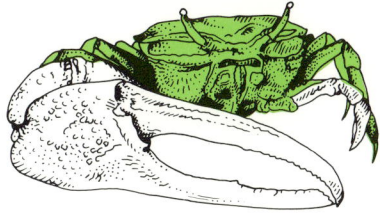

Figwort A group of square-stemmed perennial herbs related to foxgloves and snapdragons. The leaves are toothed and somewhat nettle-like and the flowers are 2-lipped and often purplish brown. They are commonly pollinated by wasps. Several species grow on river banks.

Filbert, *see* HAZEL

Finch A small seed-eating songbird found in many parts of the world. There are many different species which together form the largest of all bird families, Fringillidae. Among the finches are the BULLFINCH, CANARY, CHAFFINCH, CROSSBILL, GOLDFINCH, GREENFINCH and GROSBEAK.

Finfoot A grebe-like waterbird whose feet are lobed like those of a coot. It is a shy bird which rarely strays from the shelter of a river bank overhung with vegetation. Finfoots are found in tropical areas of South America, Africa and South-East Asia.

Fingerling, *see* SALMON

Fin whale, *see* RORQUAL

Fir Any of 40 species of evergreen conifers related to the pines. Growing in the northern hemisphere, they have flat needles and upright cones which fall to pieces when their seeds are ripe.

Firebelly This is a small toad with an unusual way of scaring its enemies. The underside is bluish-grey but spotted with red or orange. When alarmed, the animal rears up and exposes this 'firebelly'. It can also exude a strong smelling poison from its skin. The firebelly lives in eastern Europe and Russia, extending into Germany and southern Sweden.

A typical bony fish

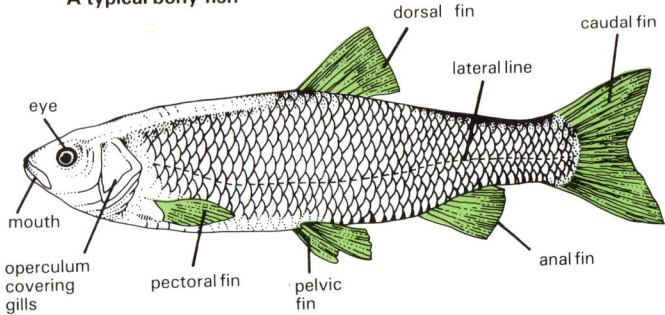

Labels: dorsal fin, caudal fin, lateral line, eye, mouth, operculum covering gills, pectoral fin, pelvic fin, anal fin

Firefly A group of winged beetles which give out a bright light from the underside of the abdomen. The light is produced by a chemical reaction. It is usually flashed at regular intervals which vary from species to species and probably serves as a recognition signal for mating. The light may be yellow, red, blue or green. The more brilliant ones live in the tropics, but several occur in the cooler regions.

Fire salamander A colourful salamander found in most parts of Europe. It is sometimes called the spotted salamander because its entire body is covered in black and yellow blotches. Like many other brightly coloured animals, the fire salamander is poisonous. The poison is secreted from pores behind the eyes and causes any animal that picks up the salamander to drop it very quickly.

Fish The name given to various vertebrate animals which live almost entirely in water and breathe through gills. Fishes are cold-blooded creatures: their bodies take on the temperature of the water surrounding them. Fishes can be grouped into freshwater and saltwater or marine forms. But some fishes can live in both kinds of water. A more scientific division is based upon body structure and gives three classes. The majority of fishes are bony fishes — they have skeletons made of bone. The cartilaginous fishes — sharks, skates and rays — have skeletons of cartilage, a gristly

substance. The third group, the lampreys and hagfishes, are jawless fishes with cartilaginous skeletons.

A fish moves through the water by bending its body to and fro and by moving its tail. The dorsal fins, along the topside of the body, and the anal fin on the underside just in front of the tail, help to keep the fish on an even keel. The two pelvic fins in the middle of the underside and the pair of pectoral fins just behind the head control vertical movements and also act as brakes. Most fishes are covered with scales and in nearly all bony fishes there is a line of special scales along the middle of the body. This is called the lateral line, and through these scales a fish can sense vibrations in the water. Bony fishes have another special organ — the swim-bladder. This is a small bladder of gas that gives the fish the buoyancy required at whatever depth it is swimming. Fishes obtain oxygen from the water by means of gills. These are feathery structures behind the head which, except in the cartilaginous fishes, are covered by a flap called the operculum. The gills are fed with blood from the heart. In the gills oxygen passes from the water into the blood.

Almost all fishes hatch from eggs. Many fishes produce millions of eggs at a single spawning, leaving the eggs and the newly-hatched fish unprotected. Other fishes are more involved with their young. Some sticklebacks build nests for their eggs and keep guard over their young. The male pipefish and the male seahorse carry the eggs in a brood-pouch on the underside of the body. Some catfishes and mouth-breeders carry the fertilized eggs in their mouths. A few fishes give birth to live young. These include several popular aquarium fishes, such as the guppy, and some sharks. Fishes are found in almost every aquatic habitat, from the coldest and deepest seas to warm springs where the temperature may reach almost 50°C. They are also the most numerous of the vertebrates, with the known species numbering more than twice that of all other vertebrates put together.

Fish-eating bat This is a more or less naked bat living in America from Mexico to Argentina. It has long, narrow wings and it flies low over the water, impaling fishes on its sharp claws. It is also called the bulldog bat.

Fisheagle Described as the most handsome of Africa's eagles, this relatively small species has a white head, back and chest, together with black wings and a chestnut belly. It lives all over the southern half of Africa and fishes along the shores of rivers and lakes. Other kinds of fisheagle live in Asia.

Flagellum A whip-like 'hair' usually found singly or paired on various single-celled organisms. The beating of the flagella enables the organism to move.

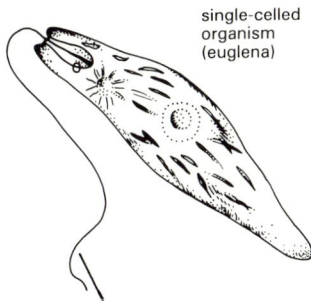

Flamingo

single-celled organism (euglena)

Flagellum

Flamingo A strange but beautiful bird with a long neck and long legs. The plumage is a delicate pink except for the flight feathers, which are black. The beak is strangely curved, and the birds feed with their heads upside-down in the water. Flamingoes live and breed in shallow lakes and lagoons. The greater flamingo, the commonest of four species, is found in many parts of South America, Africa and Central Asia. It also lives in southern Europe.

Flatworm The general name for members of the invertebrate phylum Platyhelminthes. Most have flattened, ribbon-like bodies. Free-living forms are found under stones in damp spots, in fresh water and in the sea. Others, such as the FLUKE and the TAPEWORM, live as parasites in the bodies of other animals.

Flax Any of 100 species of slender herbs of temperate lands. They grow from 30 to 120 cm tall and have small narrow leaves and most have conspicuous blue flowers, though some have white or yellow flowers. The best known species is the annual *Linum usitatissimum*, whose stems yield the fibres which make linen. Its seeds also yield linseed oil.

Flea The name given to more than 1000 species of small, wingless insects, all of which are blood-sucking parasites of mammals and birds. Some fleas have only one species of host. Others are equally at home on a variety of animals. The dreaded bubonic plague, or 'black death', which swept across Europe in the Middle Ages was carried in the saliva of fleas which normally live on rats but infest human beings when their original host dies. Flea larvae are not blood-suckers. They feed on animal and plant debris.

Fleabane Any of 150 species of composite herbs of the northern hemisphere. They have yellow, blue or white daisy-like flowers. Drugs are made from some species. The common European species was believed to banish fleas from houses when strewn on floors or burnt in fireplaces.

Flicker Admired for their beautiful plumage and wide variety of calls, the flickers are small American woodpeckers found from Alaska to Chile. Most of them feed on ants, which they collect on tree-trunks or on the ground, often tearing open the nests to get at the insects. .

Floret One of the elements of a composite flower such as a dandelion. Each floret is actually a complete flower in that it has its own sex-organs, calyx and corolla. The strap-shaped part of the dandelion floret and of the outer florets of daisies is called the ligule.

Flounder A flatfish of the shallow seas around the Atlantic Ocean, the flounder may reach a weight of 2.5 kg. It is at home in both salt and fresh water and it frequents river mouths. It is greyish-brown on the upper side and pearly white underneath.

Flower That part of the flowering plant that is concerned with reproduction. A flower is made up of several parts, all of which can be seen in a flower such as the buttercup. The stalk of each flower is called the pedicel and is swollen at the tip, forming the receptacle. The floral organs develop in more or less concentric circles on the receptacle. The first-formed organs are the sepals, of which there are five in the buttercup. They are green and leaf-like and together form the calyx whose chief function is the protection of the developing flower. Above the sepals are the five yellow petals, each with a small

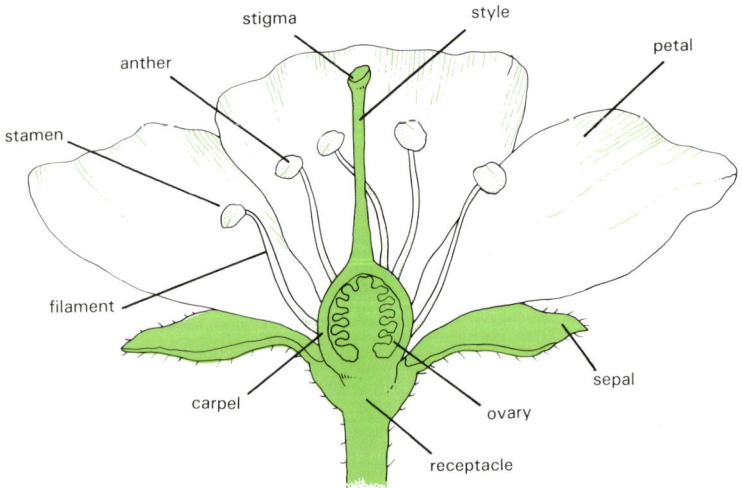

The parts of a flower

nectary at the base. The group of petals is called the corolla and, with the calyx, forms the perianth.

The petals and nectaries attract pollinating insects to the flowers and also help to protect the essential organs within. These essential organs are the stamens and carpels. The stamens are the male, pollen-producing organs and form the androecium. Each stamen consists of a filament and anthers. The carpels. the female organs, each contain an ovule and form the gynaecium in the centre of the flower. The stigma is the tip of the carpel through which the pollen tube from the pollen grain gains access to the ovule. The buttercup has many stamens and carpels but this is not so in all flowers. Below the flower there may be one or two tiny leaves called bracteoles.

The buttercup has all four types of floral organ — sepals, petals, stamens and carpels — arranged in a regular manner and is a so-called 'perfect flower'. There are, however, many variations of this pattern. Not all plants have bisexual flowers — the hazel for example has separate male and female flowers with only stamens or carpels. The flowers of peas and many orchids are highly irregular and have oddly-shaped petals. Petals (and sepals) often join and form a tube (e.g. primrose, bluebell) or may even be absent — grass flowers have neither petals nor sepals. These are just a few of many variations.

Flowering rush A perennial marsh herb of Europe and Asia. It has tall slender leaves and rose-red flowers.

Flowerpecker A group of tiny, sparrow-like birds living in forests of south-eastern Asia and Australia. Most flowerpeckers feed on the nectar and berries of plants such as mistletoe. In doing so, these birds not only help in pollinating mistletoe flowers; they also help to distribute mistletoe seeds.

Fluke A group of parasitic flatworms. Some flukes live in the bloodstream. Many have complex life histories involving two different kinds of host animal. Best known is the liver fluke which infects the livers of domestic animals, particularly sheep, in which it causes a disease called rot.

Fly An insect belonging to the order Diptera. Flies differ from other insects in having only one pair of wings (Diptera means two-winged). The hindwings have become small club-like growths called halteres which are used as balancers in flight. These insects are sometimes called 'true flies' to distinguish them from other insects with 'fly' as part of their name. The true flies include MOSQUITOES, CRANE-FLIES, HOUSE-FLIES, HOVER-FLIES and HORSE-FLIES among many others.

Flycatcher A group of small insect-eating birds found in most parts of the Old World. Flycatchers are woodland birds and most are good songsters, especially those in Australia known as robins.

Flying fish There are two types of flying fish. One has only one pair of wings — formed from the pectoral fins — and the other has two pairs of wings, formed from both pectoral and pelvic fins. Both live in the tropical seas and feed on plankton. As they break the surface, the fins are spread and, with a few flicks of the tail, they take off in a glide which may well last for over half a minute, but they do not flap their fins in flight.

Flying fish

Flying fox The name given to a group of large fruit-eating bats with fox-like faces. Unlike the insect-eating bats, they depend upon their large eyes to find their way about. They range from India to Australia.

Flying lemur Despite its name this animal is not related to the lemurs, though it resembles them in appearance. The two species of flying lemur are placed in a separate order of mammals between the insectivores and the bats. They live in the forests of South-East Asia and rarely descend to the ground. Equipped with a membrane of skin on each side of the body running from the chin to the tip of the tail, flying lemurs are able to make gliding leaps of 100 metres or more from tree to tree.

Flying phalanger A group of squirrel-like marsupials living in Australia and New Guinea. All are tree-dwellers and glide from tree to tree using a hairy web of skin stretching between their front and back legs as a parachute. 'Flights' of 30 metres or more have been recorded.

Flying squirrel A group of squirrels which have a furry flap of skin stretching along the sides of the body from front legs to back legs. The squirrels glide from tree to tree on these outstretched flaps of skin. Most flying squirrels live in southern Asia, but one species lives in North America and one in northern Eurasia. All are nocturnal animals.

Foetus The mammalian embryo when all the main features are recognizable.

Food chain A fly sucks plant nectar; a spider eats the fly; a small bird eats the spider; a bird of prey eats the small bird. This is a typical food chain. Material passes from the plant, through several animals before it reaches the last link in the chain. The cycle is completed with the death and decay of the last animal in the chain. The simple substances formed by decay will be used again by a plant and a new cycle will start.

Foraminifera Tiny marine protozoans related to the AMOEBA which build shells from lime salts in sea water. Foraminifera shells cover large parts of the ocean floor and are the principal ingredient of chalk rocks.

Foxes are renowned for their cunning, a characteristic which has enabled them to survive despite being frequently hunted for sport.

Flying lemur

Forget-me-not A small European herb related to BORAGE. It has small, blue (sometimes pink) flowers and hairy lance-shaped leaves. There are several very similar species.

Forsythia Any of a group of deciduous shrubs of the olive family, native to China and Japan. They have spreading branches, long jagged leaves and clusters of yellow flowers, which generally appear before the leaves open.

Fossa A cat-like carnivore from Malagasy (Madagascar), related to the civets. It has short soft, reddish-brown fur and it measures about 2.5 metres in length, half of which is the stout tail. The fossa is a nocturnal animal, feeding on lemurs which it catches in the trees.

Four-eyed fish A minnow-like fish found in rivers of South and Central America. It is remarkable in having a pair of protruding eyes divided into two parts, like a pair of bifocal spectacles. The fish swims just beneath the surface so that the top half of each eye is above the water and the bottom half beneath. This arrangement may mean that the four-eyed fish can search for the water animals on which it feeds while keeping a lookout for enemies from above, such as water birds.

Fox A predatory member of the dog family commonly found in most parts of the world. Among the best known species are the European RED FOX and its close North American relative bearing the same name, the GREY FOX of North and South America, the FENNEC FOX of northern Africa and Arabia and the ARCTIC FOX.

Foxglove The name of a group of perennial and biennial plants related to figworts. Foxgloves have tall spikes of flowers, each shaped like a finger-stall. The blooms are purple, yellow or white. The poison digitalis, used as a heart drug, comes from foxgloves.

Fox shark, *see* THRESHER SHARK

Francolin A group of ground-living birds found in Africa and Asia. They belong to the pheasant family and are like large partridges weighing up to 1.5 kg.

Freesia A South African plant growing from a corm, related to irises. It has a very strong sweet scent which is the hallmark of the freesia. The long leaves are sword-like and the funnel-shaped flowers are white, yellow, orange, red or purple.

Frigatebird Also known as man-o'-war birds, frigatebirds live around the coasts in the warmer parts of the world. They are not really sea-birds, for their feet have only traces of webbing and their plumage is not very waterproof.

The plumage is mainly black, shot with blue or green, and the males have a red throat pouch. The wings are large, spanning 2 metres in some species, and with a body weight of only 1.5 kg they are able to glide and soar in the slightest breeze. A frigatebird will rob smaller sea-birds of their food by attacking and tormenting them until they drop the food, which the frigatebird will swoop down and catch before it hits the sea and sinks. It is this aggressive behaviour that has earned them the alternative name of man-o'-war birds.

Frilled lizard This is a pale brown lizard, sometimes bearing yellowish or dark patches, named from the large frill around its throat. The frill normally lies like a cape around the shoulders, but it can be raised as a warning display when the animal is disturbed. The frilled lizard reaches a length of about one metre. It lives mainly in the sandy, semi-dry areas of northern Australia and it feeds on insects and small mammals.

Frilled shark A rare, eel-like shark about 2 metres long which lives in the Pacific Ocean. The six pairs of frilly

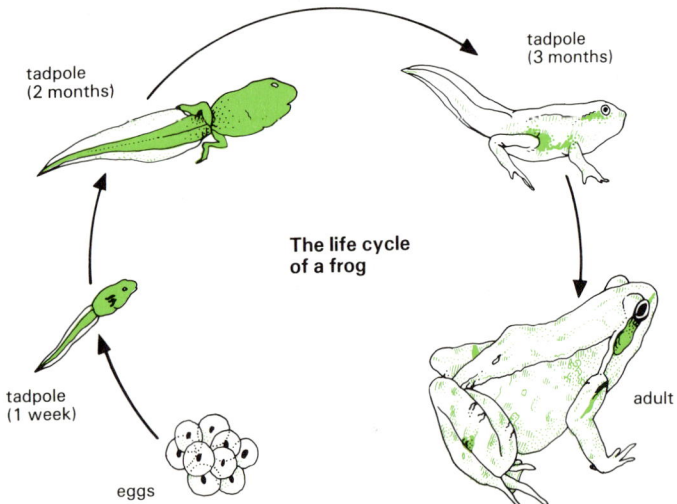

tadpole
(2 months)

tadpole
(3 months)

**The life cycle
of a frog**

tadpole
(1 week)

adult

eggs

gills which give this fish its name are like those found in fossils of the earliest known sharks.

Fritillary (botanical) Any of about 100 species of bulbous herbs related to lilies. Fritillaries have nodding, bell-shaped flowers and grass-like leaves. The common fritillary is sometimes called snake's head from the shape of the unopened bud. Its flowers are generally mottled purple and white. Other species have yellow or orange flowers.

Fritillary (zoological) This is the name given to a group of butterflies which have a chequered pattern rather like that of the fritillary flowers. Most of the fritillaries have orange-brown wings marked with black, and the undersides often bear silvery spots. The majority of the species live in woodland or on rough grassland.

Frog The frogs and toads are amphibians making up the order Salientia. They use their long, webbed hindlegs for both jumping and swimming. The smooth-skinned species tend to be called frogs and the rough-skinned species toads, though this is not a scientific division. Some frogs and toads live permanently in the water. Others live permanently on land. But the vast majority lead lives like that of the common frog, where the adults spend most of their time on the land and return to the water only for a short time during the breeding season. The frogs pair up in the water and the male fertilizes the eggs as soon as the female has laid them. The jelly-coated egg mass is known as frogspawn. The eggs develop into limbless larvae called tadpoles which breathe through feathery external gills and swim through the water with the aid of their long tails. During development the gills are replaced by internal gills, rather like those of fishes, and finally by air-breathing lungs. Meanwhile the hindlegs and then the forelegs form and the tail begins to disappear. Eventually the tadpole becomes a miniature frog and leaves the water.

Frogbit A floating aquatic herb of Europe and northern Asia. It is found in ditches and ponds, and has kidney-shaped leaves on long stalks and white, three-petalled flowers.

Frogmouth A group of nocturnal birds living in the tropical forests of South-East Asia and Australia. The name refers to their wide mouths which open in a frog-like gape.

Fructose, *see* CARBOHYDRATE

Fruit A feature of flowering plants which is formed from the carpel(s) and which helps to protect and distribute the seeds. Fruits which are derived from the carpel(s) alone are called true fruits: those that contain parts of other organs as well are false fruits. True fruits include the DRUPE (e.g. plum), the BERRY (e.g. tomato), DEHISCENT fruits (e.g. pea pod) and INDEHISCENT fruits (e.g. acorn). False fruits include the POME (e.g. apple).

Fruits

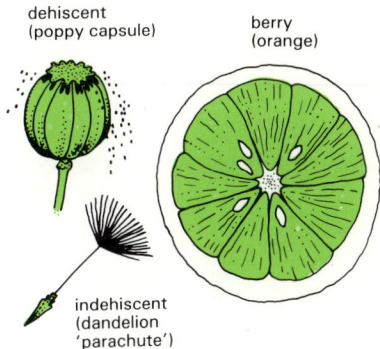

dehiscent
(poppy capsule)

berry
(orange)

indehiscent
(dandelion
'parachute')

Fruit bat The fruit bats live in the warmer parts of the Old World. About 60 of them are known as flying foxes. They vary in size from 5 cm in length to large bats with wingspans of almost one metre. Although called fruit bats, not all of them actually eat fruit. They differ from insect-eating bats in several ways, notably in the complete or almost complete absence of a tail.

Fruit-fly A group of small flies, about 2.5 mm in length. They are pests in kitchens and bars where they feed on vinegar and alcoholic drinks. But they are important to biologists studying HEREDITY. Their value lies partly in their unusually large CHROMOSOMES, which can be studied easily under a microscope, and partly in their rapid breeding rate of only two weeks from one generation to the next.

Fuchsia Any of 70 species of Central American flowering shrubs, now wild and cultivated elsewhere. Fuchsias have funnel-shaped hanging flowers, usually some shade of red. In their native homes fuchsias are generally pollinated by birds.

Fulmar A member of the petrel family, the fulmar's name is derived from 'foul bird' — a name based on its habit of spitting a foul-smelling oil at intruders. The bird has a stocky body and long narrow wings. The fulmar is one of the commonest birds in the North Atlantic.

Fungus (plural fungi) Any of a group of simple plants with no green colouring matter, that cannot make their own food. Fungi draw their food from dead or live plants and animals. They include MUSHROOMS, TOADSTOOLS, MILDEWS, MOULDS, RUSTS and SMUTS.

Fur seal The fur seals belong to the eared seal family, a group which also includes the sea lions. They differ from true seals in having small ear flaps and

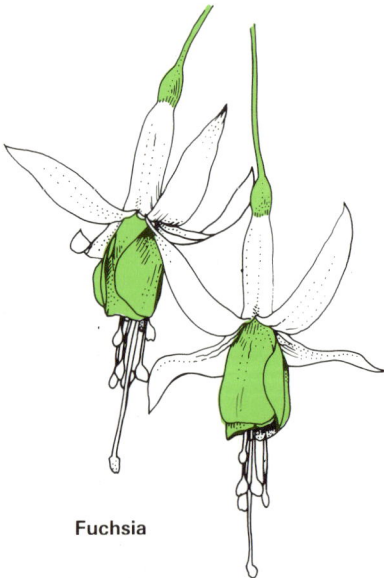

Fur seal

in being able to turn their hindlimbs forward when on land. Fur seals also have a dense fur. Most of the fur seals live in the southern hemisphere, but one species lives around the shores of the North Pacific.

Furze, *see* GORSE

G

Gaboon viper, *see* PUFF ADDER

Galagos, *see* BUSHBABY

Gall An abnormal plant growth produced in response to infection with a parasite. Examples include the robin's pin-cushion and oak-apples caused by infection with insect larvae; potato wart and other swellings caused by fungi; and even the swelling caused by mistletoe growing on an apple branch.

Gall-bladder A thin-walled sac in the liver which collects bile. When food enters the intestine the gall-bladder contracts and bile passes to the intestine.

Gallinule A group of birds related to the coots. They live near water, in marshes or by streams and ponds. The purple gallinule or swamp hen is found in southern Europe, Africa, Asia and Australia. The American gallinule lives in the warmer parts of North and South America.

Gall-wasp The familiar oak-apples and marble galls on oak trees, and the robin's pin-cushion on wild rose bushes are caused by insect larvae.

Fuchsia

These various swellings are known as galls. They can be caused by various kinds of insect larvae, but a large number of them are caused by the larvae of gall-wasps. The gall-wasps are little ant-like creatures, only distantly related to the true wasps.

Game pheasant, *see* PHEASANT

Gamete A sex-cell, i.e. a sperm or ovum (egg-cell).

Gametophyte The generation of plants such as ferns or mosses that bears sex-cells or gametes.

Ganglion (plural ganglia) A swelling in a NERVE which contains the actual cell bodies and nuclei of the nerve cells or neurons. The rest of the nerve contains bundles of fibres which actually carry the neurons' signals from one part of the body to another. Each fibre is, of course, connected to a cell body in one of the numerous ganglia. The brains of some simple animals, such as earthworms, are little more than large ganglia.

Gannet White, goose-sized birds that live around the temperate oceans of the world. They are oceanic birds, coming ashore only in the breeding season. They feed on fish and will swim after their prey under water. Gannets nest on steep cliffs and offshore rocks, and their nests are often packed tightly together.

Gaper A bivalve mollusc named for the fact that it cannot completely close the two halves of its shell. The gaper buries itself in muddy seashores and extracts oxygen and food particles from the water when the tide is in. The sand gaper is one of the best known species. In America it is the basis of clam-bakes and chowders.

Gar Also called gar-pike, these fishes live in rivers and lakes of North and Central America. They are rather lazy fishes, but they can move very rapidly to snap up an unwary fish in their long toothed jaws. Gars have rather slender bodies ranging from 30 cm to over 3 metres in length. The body is covered with tough diamond-shaped scales like those of the early fishes.

Garlic A pungent bulbous plant related to the ONION. It was originally wild in Central Asia. Unlike the onion, each garlic 'bulb' is actually a cluster of small wedge-like bulbs called cloves. Garlic is used as a flavouring. The plant has a tall stalk with whitish flowers.

Garter snake The commonest snakes in North America. There are many species, none of which is poisonous. The body is slender and marked with longitudinal stripes. Worms are the major items in the garter snake's diet.

Gasteropod A member of the mollusc group Gasteropoda which includes snails and slugs. One feature they have in common is a muscular foot used to move about. Another is the possession of a shell, though in slugs this is almost or entirely absent.

Gastric Concerning the stomach.

Gavial, *see* GHARIAL

Gazelle Dainty, slender antelopes in which the males have sweeping horns that diverge from each other to form a lyre-shaped outline. The females have only short spikes or no horns at all. There are 10 species of true gazelle, most of them living in Africa. Grant's gazelle and Thomson's gazelle are among the best known of the African species.

Gecko

Gecko The geckos form a family of lizards noted for the broad toes of many of its members and for the microscopic hooks on the toes. These can hook into all but the smoothest of surfaces, and so the geckos can climb up almost anything. Geckos live in all warm countries and there are many species. They vary from about 2.5–35 cm in length.

Gemma A small reproductive body that forms and breaks away from mosses and liverworts and grows directly into a new plant.

Gemsbok

Gerbil

Gemsbok, *see* ORYX

Gene An hereditary factor passed from generation to generation of plants and animals. Genes are carried on the chromosomes in the cell nuclei. They are believed to be composed of nucleic acids which control the development of the cells by controlling the types of protein formed. The vast number of instructions carried by the genes of an organism is made possible by the enormous variation in the arrangement of the molecules making up the nucleic acids.

Genet The genet looks like a cross between a tabby cat and a mongoose. It is a carnivorous mammal related to the mongooses and civets. There are six species, all but one confined to Africa. They feed on small rodents, birds and insects.

Genetics The study of GENES and the way in which characteristics are passed on to offspring.

Gentian The name of about 600 species of herbs, mostly with funnel-shaped flowers. Many gentians have vivid blue blooms; others are purple or yellow and some are white. Bottle gentians have flowers that do not open. Most gentians are mountain flowers.

Genus A category used in classification, consisting of a number of closely related species, all of which share the generic name. For example, the leopard and the jaguar both belong to the genus *Panthera* and are scientifically known as *Panthera pardus* and *Panthera onca* respectively.

Geranium A group of low, ornamental herbs with small regular flowers and often with divided leaves. They are found all over the world. The flowers are shades of red and blue. Some cultivated forms are known as pelargoniums.

Gerbil Also known as sand rats, gerbils live in the arid parts of Africa and Asia. The fur is fawn or brown, often tinged with black, and the underside is pale. The tail is usually long and the animals also have long back legs. Most of the species can jump. Gerbils eat seeds, roots, and other parts of plants. They will also eat insects.

Gerenuk A long-necked, gazelle-like antelope. The male carries short, thick horns and stands about one metre high at the shoulder. Gerenuks feed by browsing on trees and they often stand up on their hindlegs. These animals live only in Somalia and neighbouring parts of north-eastern Africa.

Germ cell A gamete or sex-cell.

Germination The germination of a seed requires warmth and moisture. The seed absorbs water and swells up until the seed-coat bursts. First to emerge is the root or radicle followed by the shoot or plumule. The young plant relies upon the food stored in the seed until the leaves open. The plant can then make its own food.

Gestation period The period from fertilization to birth in mammals. It varies from about three weeks in house mice, to about 40 weeks in human beings and nearly 2 years in elephants.

Gharial A long, slender-snouted crocodilian living in the rivers of northern India and Pakistan. It is also known as the gavial. It can reach a length of about 6 metres and keeps to the water more than other crocodiles.

Gherkin, *see* CUCUMBER

Ghost bat, *see* TOMB BAT

Ghost moth This night-flying moth earns its name by the way that the male vanishes and reappears during flight. This is due to its wings being dark underneath and white above. The females are very different in appearance with brownish hindwings and yellow forewings. Ghost moths are found in many parts of Europe and the Middle East.

Giant crab, *see* SPIDER CRAB

Giant panda, *see* PANDA

Gibbon The smallest of the apes, gibbons are among the most agile of mammals. They are normally about one metre in height when standing upright. The arms are extremely long, perhaps half as long again as the legs, and they are used for swinging through the trees. Gibbons live in the forests of south-eastern Asia and feed mainly on fruit. There are several species.

Gila monster This is a lizard living in the deserts of the south-western United States. It is up to about 60 cm long and it is one of only two poisonous lizards. The other is the similar looking beaded lizard which lives in Mexico. The gila monster is pink and yellow, with black markings. It is a rather sluggish animal, feeding on birds' eggs and young mammals.

Gill The organ which allows fishes and many other aquatic animals to breathe underwater by absorbing oxygen dissolved in the water. The gills of fishes consist of small finger-like organs connected to the throat and opening to the outside of the body through gill slits.

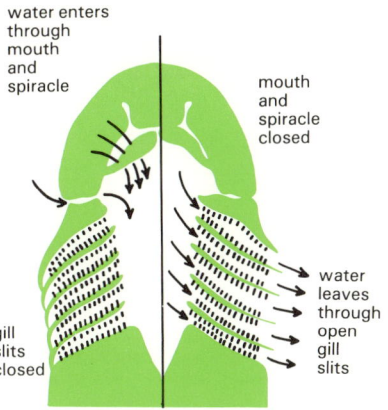

Gills of a shark

The gills of crabs and lobsters are feathery outgrowths from the body.

Ginger A plant of the East Indies, widely cultivated in tropical regions for the spicy flavour of its rhizomes. It has lance-shaped, narrow leaves and white and purple flowers.

Ginkgo (or maidenhair tree) A slender tree with heart-shaped leaves. It is the only survivor of a group of plants that flourished millions of years ago. It is now native to China but widely grown elsewhere in parks and gardens. The

Germination of a bean

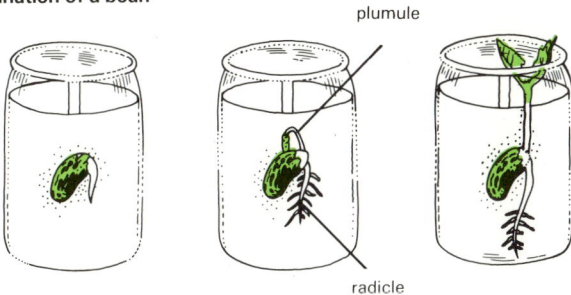

It is possible to watch a bean germinate by placing it in a jar lined with damp blotting paper. First the root, or radicle, will begin to grow downwards, then the plumule will grow upwards.

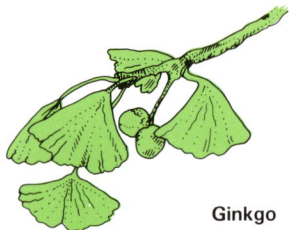

Ginkgo

female tree bears unpleasant-smelling fruit while male trees bear short catkin-like cones in spring.

Ginseng Two species of herb whose roots are used as a drug with alleged rejuvenating powers. One grows in eastern Asia, the other in North America. It is a low-growing plant whose leaves each consist of three or more leaflets. It has greenish flowers.

Giraffe The tallest animal in the world, the giraffe has remarkably long legs and a remarkably long neck. An old male may reach 5.5 metres in height. Giraffes live in the savanna areas of Africa, mainly in the east and south. They browse on the tree-top foliage with their long tongue and mobile, hairy lips.

Giraffe

Gizzard A muscular region of the alimentary canal of certain animals where food is ground up prior to main digestive processes, especially well-developed in seed-eating birds.

Gladiolus The name of a group of flowering plants with sword-shaped leaves. Gladioli are related to IRISES. They grow from corms and have spikes of colourful, tubular blooms. Most come from South Africa.

Gland An organ that manufactures certain substances and passes them out into the body where they fulfil particular functions. Digestive juices are produced in glands such as the salivary glands and the pancreas. These have ducts to carry away the secretions and are called exocrine glands. There are, however, several ductless or endocrine glands, such as the thyroid and adrenals. These produce hormones which are carried to their sites of action by the blood.

Glass snake Legless lizards found in many parts of the world. They reach lengths of up to one metre, but two-thirds of this length is tail. Glass snakes live in open country, hiding under leaves and stones. They feed mainly on insects. The fragile tail is easily broken off if seized — hence the name of glass snake.

Gliding frog The gliding frogs belong to a family of tree frogs and they get their name from their habit of gliding from tree to tree, supported by the webbing of the large feet. The most common gliding frogs live in Malaya and Borneo.

Glis-glis, *see* DORMOUSE

Globe artichoke, *see* ARTICHOKE, GLOBE

Glow-worm The glow-worm is a beetle belonging to the same family as the fire-flies. The female is wingless and it is she who is responsible for the name glow-worm. All stages of the life-cycle can give out light, but the female shines most brightly. She climbs up a grass stem and shines to attract the flying males. Young glow-worms look very much like the females.

Gloxinia A plant with dark green velvety leaves and rich purple flowers. It is a native of Brazil and is a popular house-plant in northern regions. The

The human endocrine glands

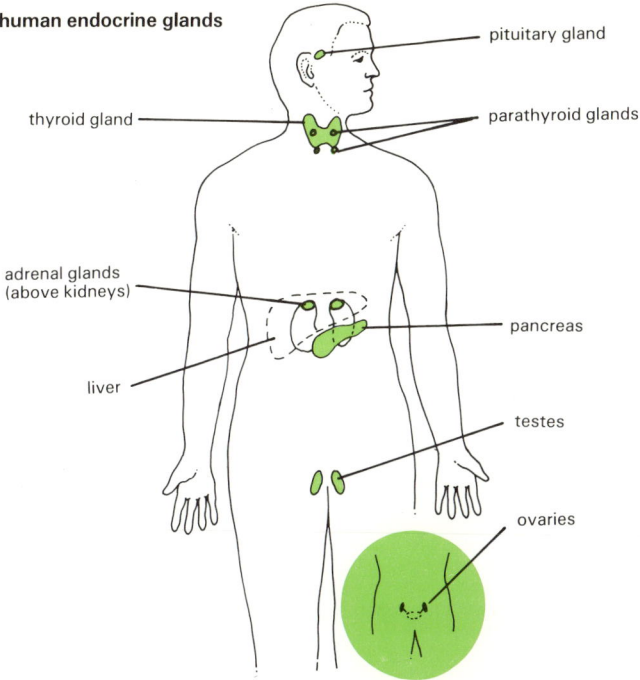

pituitary gland

thyroid gland

parathyroid glands

adrenal glands
(above kidneys)

pancreas

liver

testes

ovaries

gloxinia is a very tender plant, growing only in gardens in frost-free regions in the summer.

Glucose, *see* CARBOHYDRATE

Glume One of the chaffy bracts that enclose grass flowers.

Glutton, *see* WOLVERINE

Glycogen Animal starch — a polysaccharide that is stored by various animals in their bodies and which can be converted into glucose when required for respiration.

Gnu Strange, cow-like antelopes of eastern and southern Africa. A fully-grown male may stand up to one metre at the shoulder and weigh 200 kg. Gnus roam the plains in search of young grass and may cover up to 50 km per day. They are also known as wildebeest.

Goat The wild goat from which the domesticated goat is descended ranges from southern Europe to western Asia. Goats are expert climbers, equally at home among rocky crags or in the branches of trees. They feed on almost any vegetable matter, from grass and shrubs to the bark and leaves of trees. Other wild goats include the IBEX and the MARKHOR.

Gnu

101

Goat-antelope

Golden eagle

Goat-antelope A group of hollow-horned ruminants which have similarities with antelopes and goats. All live in mountainous areas. The CHAMOIS of Europe and the ROCKY MOUNTAIN GOAT of North America are two examples of goat-antelope.

Goat moth A large drab moth with mottled wings spanning 7 cm or more. The name refers to the goat-like smell of the larvae which feed on wood and burrow into trees, especially willows and poplars.

Goatsbeard A Eurasian herb with long narrow leaves and dandelion-like flowers. It has a long tap-root and milky juice. Salsify, a root vegetable, is a kind of goatsbeard.

Goatsucker, *see* NIGHTJAR

Goby A small group of colourful marine fishes with flattened heads and large eyes. Most are less than 10 cm in length, the largest being the MUDSKIPPERS.

Godwit A large wading bird related to the snipe. Godwits have long legs and a long bill, slightly upturned in some species. The two European species, the black-tailed godwit and the bar-tailed godwit, migrate far to the south in the winter, the bar-tailed godwit occasionally reaching Australia.

Goldcrest Once known as the gold-crested wren, this is the smallest British bird, weighing about 5 grams or half the weight of a wren. It is an insect-eating member of the warbler family found in most parts of Europe and ranging across Asia to Japan. Goldcrests, named for the yellowish-orange patch on the head, are particularly at home in coniferous woods. The firecrests of North America and parts of Europe are closely related to the goldcrest but have an orange patch and black and white stripes over the eye.

Golden eagle Inhabiting mountainous countryside throughout most of the northern hemisphere, the golden eagle is a majestic bird reaching almost one metre in length with a wingspan up to 2.5 metres. The diet of the golden eagle varies throughout its range, but it consists largely of rodents, rabbits and ground-nesting birds. Larger animals, such as lambs or deer calves, are usually eaten only as carrion.

Golden oriole, *see* ORIOLE

Golden rod An erect perennial herb of Europe and North America, bearing clusters of small yellow composite flower-heads. It has narrow leaves. It was once used for treating wounds.

Goldfinch A colourful bird named for the golden bar on each wing. It is found in Europe, south-western Asia and North America. Goldfinches live in small flocks except in the breeding season. They are mainly seed-eating birds.

Goldfish An ornamental fish derived from a carp native to China. The wild form is usually greenish brown in colour but occasionally an orange individual is produced, and it is these which have been bred for ornamental purposes. Goldfish have been known to live for at least 25 years in captivity, but their life-span in the wild is far less.

Gonad An organ where sex-cells are formed — i.e. testis or ovary.

Good King Henry, *see* GOOSEFOOT

Goose A large semi-aquatic bird belonging to the family Anatidae which also includes the ducks and swans. Examples include the BARNACLE GOOSE, the CANADA GOOSE, the GREY-LAG GOOSE and the HAWAIIAN GOOSE.

Gooseberry A small shrub, closely related to CURRANTS, producing tart green fruits. It has spiny stems, and small toothed leaves. It is widely cultivated, but is native to Europe and North America.

Goosefoot A group of plants found world-wide, so called from the shape of the leaves. Other names are lamb's quarters and pigweed. One species, Good King Henry, was formerly eaten as a substitute for asparagus. Mexican tea is another species, used as a tonic. All have inconspicuous green flowers without petals.

Goral A goat-antelope with a long coat and a bushy tail. Gorals are found in the western parts of the Himalayas living at altitudes up to 3000 metres.

Gorilla The largest of the man-like apes, the male reaches almost 2 metres in height and weighs around 200 kg. The hair varies in colour from brownish black to jet black. Gorillas normally walk on all fours, in a semi-upright posture because their arms are longer than their legs. There are several races of gorilla found in different parts of central Africa. They are peaceable creatures, living in troops and feeding largely on fruit, especially bananas.

Gorse A spiny European shrub, also called furze or whin. The spines are modified evergreen leaves. The shrub belongs to the pea family and bears bright yellow flowers at most times of the year, but especially in spring.

Goshawk A magnificent bird of prey, the goshawk is closely related to the sparrow-hawk. It has the same long tail and rounded wings when seen in silhouette, but it is much larger — the head and body measuring about 60 cm. The plumage is dark brown, with a pale underside marked with bars. The goshawk ranges over much of the higher latitudes of the northern hemisphere. It is mainly a woodland bird.

Goral

Gouldian finch Rainbow finch and painted finch are alternative names for this handsome bird of northern Australia. The plumage of the adult male is brilliantly patterned with scarlet, blue, black, lilac, green and yellow. Females are less colourful and young birds are green overall.

Gourd An annual trailing plant of Africa, southern Asia and America related to marrows and squashes. The many species bear large thick-skinned fruits, often highly ornamental. Some can be eaten; other fruits are dried and the skins (shells) are used as dishes or flasks.

Grackle A group of birds resembling starlings both in shape and behaviour which are fast becoming pests of North American towns and farms. Like starlings, they live in large flocks and feed on a wide variety of food ranging from grain to the nestlings of other birds. Grackles have black plumage with an iridescent sheen.

Grafting, *see* PROPAGATION

Grant's gazelle, *see* GAZELLE

Grape A woody vine of the north temperate zone, bearing bunches of juicy berries. The leaves are large, lobed and with toothed edges. Cultivated on a very large scale, the vines do best in dry, warm regions. Grapes are used for wine, to dry as raisins, sultanas and currants, or just to eat.

Grapefruit A large citrus fruit, related to ORANGES, growing on trees in warm climates. The tree has dark green shiny leaves. Grapefruit probably originated in the West Indies.

Grass The grasses are members of one of the largest plant families — the Gramineae. They grow almost everywhere. They have jointed stems, and long slender leaves which grow from the base. The tiny dull flowers are surrounded by papery scales (chaff) to form spikelets. Grasses include CEREALS.

Grasshopper A group of jumping insects which live among the grass and herbage close to the ground. They are

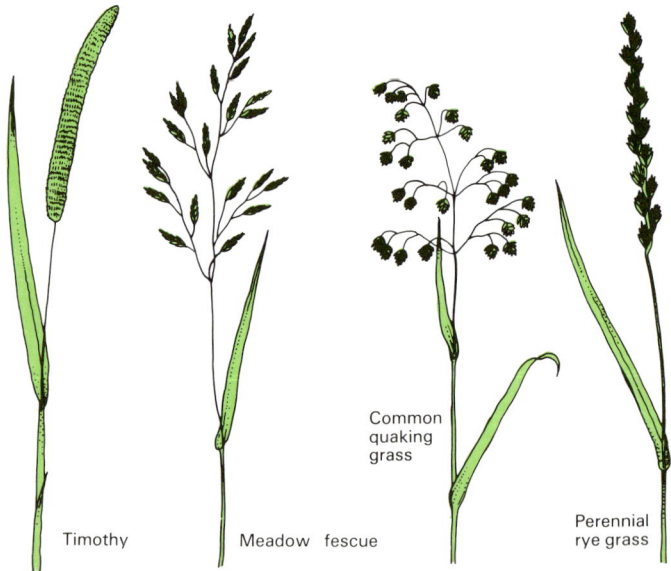

Common quaking grass

Timothy

Meadow fescue

Perennial rye grass

A variety of grasses

Grasshopper

active only when the sun is shining, and their usual method of getting about is to jump with the aid of the long and powerful back legs. Most of the species have wings and they can fly. They are best known for their 'songs', which are produced by rubbing the hindlegs against prominent veins on the wings. Usually only the male sings.

Grass snake The common grass snake extends from the British Isles (not Ireland) across Europe to central Asia. It is a non-poisonous snake reaching a length of 75–90 cm as a rule, although specimens twice this size have been

found on heathland and in dry grassy places.

Greater kudu, *see* KUDU

Grebe A group of long-necked water birds, many of which have plumes on the head. The feet are not webbed, but the toes have horny fringes that act as paddles. They live on lakes and slow-moving rivers in many parts of the world. Grebes do not fly much, although some species migrate to the coasts for the winter.

Green algae A group of simple flowerless plants growing in fresh or salt water. Most freshwater species are microscopic, but can colour a whole pond green. A number of green algae grow on damp tree-trunks, fences, etc. The larger plants are seaweeds, found on shores throughout the world.

Greenbottle, *see* BLOW-FLY

Greenfinch Also known as the green linnet, the greenfinch ranges across Europe and much of Asia. It is a seed-eating garden bird about 15 cm long. The greenfinch's brownish plumage is

Great crested grebe

recorded. The snake is olive brown or a greenish colour and it can be recognized by the yellowish collar around the neck. It lives in damp places and is a good swimmer. Worms, slugs, frogs and small mammals are the main items on the grass snake's menu.

Grayling A freshwater fish related to the salmon found in most parts of northern and central Europe. It is a favourite fish among anglers and is as tasty to eat as trout. Grayling may reach a weight of 2 kg or more. The name is also given to a brown butterfly

tinged with green for most of the year, and the wings are edged with yellow.

Greenfly, *see* APHID

Greengage A variety of PLUM.

Greenland right whale, *see* RIGHT WHALE

Green linnet, *see* GREENFINCH

Green lizard Found over much of southern Europe, the green lizard reaches a length of 40 cm including a 25-cm tail. Its most northerly range is the Channel Islands.

Green mamba, *see* MAMBA

Green monkey, *see* GUENON

Green turtle Once abundant in tropical seas, the green or edible turtle is much rarer now because its flesh and eggs are good to eat. The adult turtle has a dark brown shell marbled with yellow. It reaches a maximum length of one metre. The name green turtle comes from the green tinge of its fat. Like all marine turtles it comes on to the shore to lay its eggs in the sand. Young green turtles are carnivorous, but the adults feed only on the vegetation growing in the shallow water.

Grevy's zebra, *see* ZEBRA

Grey fox Found in most parts of the United States and Central America, the grey fox is especially at home in forested areas. It climbs trees skilfully when pursued or simply in search of food. A fully-grown male may measure over one metre in length including a long bushy tail and weigh more than 70 kg.

Grey mullet, *see* MULLET

Greylag goose This large grey goose, weighing up to 4 kg, is believed to be the ancestor of the farmyard goose. It breeds in northern and eastern Europe, migrating southwards in the winter.

Grey partridge, *see* PARTRIDGE

Grey seal This seal is found on both sides of the North Atlantic. The largest colonies by far are around Scotland where these seals are regarded as pests by fishermen because they take salmon from nets. Males can reach almost 3 metres in length and weigh 275 kg. The females are slightly smaller.

Grey squirrel, *see* SQUIRREL

Grey whale A slow-swimming baleen whale which reaches a length of 14 metres and a weight of 20 tonnes. Grey whales are confined to the northeastern Pacific. Every year they migrate south to give birth to their young in the coastal waters of California and Mexico.

Gribble A marine crustacean resembling a tiny woodlouse. Gribbles burrow into wood and cause enormous damage to the submerged timbers of piers and jetties.

Grizzly A race of North American brown bear inhabiting the western high timberlands of North America. Grizzlies are known for their ferocious disposition, particularly when females are followed by cubs. Their diet includes fish, insects, fruits and berries, small animals and carrion. The cubs are born to the female while she rests half asleep during the coldest winter months, and remain at her side for about two years. Males are solitary animals and associate with the females only during the spring breeding season.

Grosbeak An old name given to a variety of finch-like birds with large conical beaks. They include the European hawfinch and pine grosbeak together with a number of American species.

Ground beetle A large family of beetles in which few species are good flyers. The adults are mostly long-legged, nimble predators. Some ground beetles defend themselves by squirting corrosive liquids from the hind end. The bombardier beetle does this with a tiny but audible bang. Most are nocturnal.

Groundnut, *see* PEANUT

Black grouse

Groundsel An annual composite herb of Europe and North Africa, having slender, toothed leaves. It has drooping yellow flower-heads, and each seed develops a silky parachute.

Ground squirrel Ground squirrels look much like tree squirrels, but they do not have such bushy tails. There are more than 30 species, most of them living in North America. Some live in Africa and some in eastern Europe and

Black-headed gull

Great black-backed gull

Common gull

Asia, where they are known as susliks. They live in burrows or under logs and they eat seeds, fruits, bulbs, insects and many other items. Those animals living in the northern regions hibernate, but the others remain active throughout the year.

Grouper Also known as sea bass or sea perch, groupers are heavy-bodied fishes living mainly in the tropical seas. The largest is the Queensland grouper, which reaches some 4 metres in length. Groupers have strong, needle-sharp teeth and feed on a variety of other fishes.

Grouse The grouse family contains 18 species of game bird, ranging in size from a domestic hen to a turkey. Most of them live in the northern regions and they feed mainly on leaves, shoots and berries. They include the black-grouse (blackcock), the capercaillie and the ptarmigan. One of the best-known American species is the ruffed grouse.

Grunion A small coastal fish of California remarkable for its breeding habits. At times of spring (high) tides, the grunions leap on to the beach; the females lay their eggs in the sand and the males fertilize them, then the fish wriggle back to the sea. The eggs hatch

at the next spring tide and the young grunions swim away.

Guanaco The wild relative of the domesticated LLAMA found in South America along the whole length of the Andes. The coat is reddish-brown above and white beneath. Guanacos live in small herds, each consisting of a ruling male and several females with their young.

Guard cell One of a pair of special cells that control the opening of a stoma (one of the pores of a leaf).

Guava A small tropical or sub-tropical American tree, producing egg-sized red or yellow fruit. The leaves are dark green and oval with downy backs, and the flowers are white. It is related to myrtle. The guava is widely cultivated in the tropics.

Gudgeon A small bronze-coloured fresh-water fish with a barbel at each corner of the mouth. It is found in rivers and lakes in many parts of Europe. A similar but much larger fish is called simply the barbel. It grows to one metre in length.

Guenon A term embracing a number of colourful African monkeys including the green monkey or vervet, the blue monkey, the red-tailed monkey, the Diana monkey and the talapoin.

Guillemot

Guillemot

They live in a variety of habitats, from the rain forest tree-tops to the open savanna. The green monkey is the commonest; it is found throughout most of Africa south of the Sahara.

Guillemot Guillemots are birds of the northern seas. They belong to the auk family and are related to puffins. On the water they look like ducks, but they stand upright on land and look more like penguins. There are various species, all black and white. Out of the breeding season, the birds stay at sea, diving and swimming underwater to catch fishes and crustaceans. They breed on steep cliffs and pack their nests tightly on to the ledges.

Guinea fowl A group of ground-living birds related to chickens found in tropical Africa. Like chickens, they scratch the ground for food, eating seeds and small animals. The commonest species is the tufted guinea fowl of the open plains south of the Sahara.

Guinea pig A small domesticated South American rodent. The wild form is usually called a cavy. The Incas used to rear guinea pigs for food, and the Spaniards eventually brought them to Europe. They are now kept as pets and as laboratory animals.

Guitarfish The name of this fish refers to its strange shape. The head and body are broad and flat, the tail is cylindrical. Guitarfishes glide along the seabed in shallow tropical seas, feeding mainly on crustaceans. Most species grow to about 2 metres in length.

Gulper eel A group of deep-sea fishes between 1 and 2 metres in length. They are all remarkably ugly, with large gaping jaws attached to a tiny head, a small body and a long whip-like tail.

Gum-tree, *see* EUCALYPTUS

Gundi This long-furred rodent is said to be the commonest mammal in North Africa. It is about 25 cm long and it looks rather like a sandy-coloured guinea pig. It lives in dry rocky areas, basking in the sunshine and seeking shelter only during the hottest part of the day. It also shelters from the rain, which damages its silky fur.

Guppy This small, freshwater fish from the Caribbean region is probably the commonest aquarium fish after the ordinary goldfish. The male guppy is a colourful little fish, about 2.5 cm long, but the female is rather drab and twice as long. She gives birth to live young, but frequently eats them as soon as they are born. Fish breeders have produced many attractive strains of guppy.

Gurnard This fish has a large box-like head and wing-like pectoral fins, in which the front two or three rays serve as 'feelers'. There are several gurnard species, generally living in shallow coastal waters.

Gymnure The gymnures are shrew-like insectivores related to the hedgehogs. They have thick underfur and a dense covering of coarse guard hairs. Gymnures come from south-eastern Asia and the largest, often called the moon rat, is about 60 cm long including its almost naked tail.

Gynaecium The female region of a flower — the carpel or carpels. Often called the ovary or pistil.

Gyrfalcon This is the largest of the falcons. Its home is the cold treeless tundra regions of northern Europe, Asia and North America. Other birds, particularly ptarmigan, are the main prey of the gyrfalcon.

H

Habitat A particular type of environment, such as woodland or marsh.

Haddock A close relative of the cod, the haddock, or finnan, is an important food fish found in the North Atlantic. It is greyish-brown above and it has a white belly. It is easily recognized by its three dorsal fins and by the dark patch just behind the gills. It reaches a length of about one metre.

Haemoglobin A red respiratory pigment that increases the oxygen-carrying capacity of the blood. In vertebrates haemoglobin is carried in the red corpuscles. Haemoglobin is based on iron, a lack of which in the diet leads to one form of anaemia. The haemoglobin of invertebrate animals, where present, is carried in the plasma.

Hair Fine outgrowths from the skin, characteristic of mammals. Hairs grow from tiny pits in the skin called follicles. Cells at the base of the follicle divide repeatedly and the new cells are pushed outwards. They soon die but do not break off: the chain of dead cells forms the hair. The hair colour is determined by pigments. Its greasiness is caused by the secretion of the sebaceous gland at the base of each hair. Hair traps a layer of air between itself and the body and helps to maintain an even temperature. A tiny muscle attached to each hair can make it stand up, although these muscles are very weak in human beings. Hairs have tiny nerves around the base and are sensitive to touch. Some hairs have extra nerves and are extra-sensitive. The cat's whiskers are very sensitive to touch.

Hairstreak A worldwide group of generally small butterflies related to the blues. The wings are mostly blue, brown or orange above and invariably

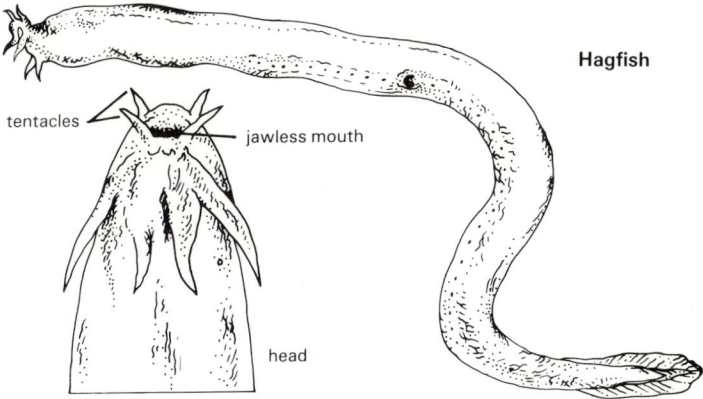

Hagfish

tentacles

jawless mouth

head

Hagfish The hagfish is a descendant of the jawless fishes that lived many millions of years ago. It bears little resemblance to other fishes except the LAMPREY. Its slimy, scaleless body is up to 60 cm long and has no paired fins. It has a large, jawless mouth, surrounded by tentacles, and a tongue carrying many horny teeth. The animal feeds by attaching itself to some other fish and rasping its flesh away.

a different colour on the underside. There is normally one or more short tails on each hindwing. The caterpillars are slug-like in shape and many have a honeygland on the back which attracts ants.

Hairy frog This is a greenish-brown frog, about 10 cm long and flushed with pink on the underside. It does not have any hairs and the name refers to the tufts of skin filaments that develop

Hake

on the sides and thighs of the male during the breeding season. These filaments are thought to increase the breathing surface of the frog. The hairy frog lives in the fast-flowing streams of West Africa.

Hake A close relative of the cod, the hake lives in the Mediterranean and the north-eastern Atlantic. It is a streamlined fish with one short and one very long dorsal fin. The hake is brownish-grey above and silvery-white on the sides and belly. It is an important food fish.

Halibut A flatfish found in the North Atlantic and North Pacific which may reach lengths of 3 metres or more and weigh 200 kg. It feeds on bottom-living invertebrates and also on other fishes. Halibut are fished for food and for the oil in their livers.

Halophyte A plant able to tolerate high salt content in the soil.

Haltere The modified hindwings of true flies (Diptera), shaped like minute pins, and used as balancing organs in flight.

Hammerhead A heron-like bird with a broad, flattened bill and a large crest on the back of its head. It is found near water over much of southern Africa.

Hammerhead shark A group of sharks in which the sides of the head are drawn out to form a hammer-head shape. They live mainly in the warmer seas and feed on bottom-living fishes and invertebrates. The largest species may reach a length of 6 metres and span one metre across the head from eye to eye.

Hamster A group of short-tailed rodents related to the voles. They range from western Europe to China and usually live in dry regions. Cereals are their staple diet and hamsters are sometimes a nuisance to farmers. The golden hamster probably came into being from a chance hybridization between two wild species in Syria. All the golden hamsters in the world have descended from a single family found there in 1930.

Haploid Having only a single set of chromosomes in the nucleus and not paired chromosomes as in the DIPLOID condition. In higher animals and plants it is only the sex cells (gametes) that are haploid.

Hare Long ears and hindlegs are the most obvious features that distinguish hares from rabbits. There are various species in Europe, Asia, Africa and North America. They include among others the brown hare and mountain hare of Europe and the snowshoe rabbit and jack rabbits of North America. Both the mountain hare and the snowshoe rabbit have white winter coats.

Harebell A perennial, low-growing herb with blue, bell-shaped flowers. It is the bluebell of Scotland. It has heart-shaped leaves near the ground, and long slender ones higher up.

Hammerhead shark

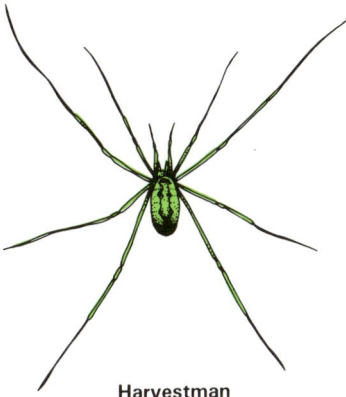

Harvestman

Harp seal A seal of the cold waters of the North Atlantic and Arctic oceans. It reaches 2 metres in length and weighs up to 200 kg. The young, with their soft white coats, are born on the sea ice between January and April and are suckled for about a fortnight. Mating takes place soon after the pups have been weaned.

Harpy eagle A large eagle with short, broad wings and powerful talons. Harpy eagles range the forests of Central and South America, swooping down to seize monkeys from the treetops.

Harrier A group of brown or blackish hawks generally living in open country. They have owl-like heads, long wings and long legs. They are found in all the continents, the hen harrier (marsh hawk in America) being the most widely distributed species.

Hartebeest A large and rather ungainly looking antelope standing up to 1.7 metres at the shoulder. The head is long and narrow and the back slopes sharply down from the shoulder to the hindquarters. Both sexes bear horns. There are three species which, between them, cover most of the southern half of Africa. They are among the commonest of the antelopes.

Harvesting ant The harvesting ants belong to various species and they get their name because they all collect the seeds of grasses and store them in their nests. They go out to forage in large bands. Small workers gather the seeds, and larger workers crush them with their jaws. Harvesting ants live mainly in warm dry regions.

Harvestman The harvestmen are related to the spiders. They are often called harvest spiders or daddy-long-legs. Harvestmen do not have the narrow waist of the spider, nor do they have the silk glands. They have four pairs of long legs, and a pair of eyes mounted on a turret on the top of the body. Most species are brownish in colour and feed mainly on other small animals.

Harvest mouse The tiny harvest mouse has a soft furry body and a long prehensile tail. Weighing only 6 or 7 grams it is light enough to be able to swing about on grass stems and make its nest there. In winter the animal lives in a burrow. The harvest mouse is found right the way across Europe and Asia. The American harvest mouse is a different creature and is really a vole.

Harvest mouse

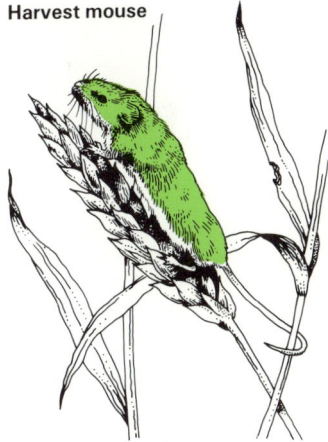

Hatchet fish A small, deep-sea fish with a shiny body flattened from side to side. This feature, together with the slender handle-like tail, is responsible for the name hatchet fish. The animals go down to 500 metres during the day, but they come nearly to the surface at night when they feed on various floating creatures.

Hawaiian goose In 1950 this relative of the Canada goose was on the verge of extinction, with less than 40 birds in all left alive. Through careful breeding in captivity that danger is now past, and many birds have been released in the wild in their native Hawaii. Also called ne-ne — their original Hawaiian name.

Hawkbit Any of a group of perennial Eurasian herbs, similar in appearance to the DANDELION. The flowers are borne on leafless stems up to 60 cm tall.

Hawkmoth Also known as sphinx moths, hawkmoths are large, thick-bodied moths found throughout the world. The wings are long and narrow and the moths can fly very fast. Most of them fly at night. The caterpillars are stout creatures and they usually have a curved horn at the hind end.

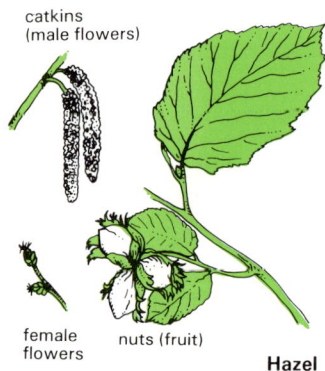

catkins
(male flowers)

female   nuts (fruit)
flowers
Hazel

Hawksbeard A composite herb with flowers and leaves somewhat like those of the dandelion. Small flowers are borne on many-branched stems. Most are perennials and some are well over one metre high. There are several species.

Hawksbill A small turtle with a hooked 'beak' found in tropical seas around the world. The hawksbill is the source of 'tortoiseshell' once widely used for ornamental purposes.

Hawkweed Any of a large group of perennial herbs of the north temperate and Arctic regions. They have small dandelion-like flowers on branching stalks, and long oval leaves, often toothed around the edges.

Hawthorn Also called quickthorn or may, the hawthorn is a thorny tree of the northern hemisphere. It has wedge-shaped, lobed leaves and clusters of white flowers. The small red fruits are called haws. The hawthorn is one of the commonest hedgerow trees, widely planted for hedges because of its quick growth and thorny branches.

Hazel Any of 15 small trees or shrubs of the north temperate zone. It has oval leaves, and its male flowers form hanging catkins very early in the year. The female flowers are small and bud-like. The edible nuts are often known as filberts or cobnuts. The strong, flexible branches can be used for basketry.

Heart This is a muscular pump which circulates the BLOOD around an animal's body. Many different designs can be found in the animal kingdom, but the most complex hearts are to be found in the birds and mammals. These animals have what is known as a double circulation and the heart consists of four separate chambers — two auricles (also known as atria — singular atrium) and two more-muscular ventricles. The blood begins its journey around the body in the left ventricle, from where it is pumped into the body's main artery — the aorta. This big artery sends branches to all parts of the body, and the finest branches or capillaries flow through every organ and tissue to distribute food and oxygen. The capillaries then join up again to form the veins, which carry the blood back to the heart. The veins flow into three major vessels, known as vena cavae, which empty the blood into the right auricle of the heart. This blood has given up virtually all of its oxygen, and there is no point in pumping it around the body again in this state. Contraction of the right auricle sends the blood into the right ventricle, and the contraction of this muscular chamber sends the blood along the pulmonary arteries to the lungs. Here it receives more oxygen and is ready to go round the body again, but first it must return to the heart. The pulmonary veins carry it to the left auricle, which pumps it to the left ventricle — and then the blood can begin another journey around the body. This is the double circulation: heart-body-heart and then heart-lungs-heart.

Reptiles and amphibians also have a double circulation, but it is generally much less efficient. There is only one ventricle, although it may be divided to a greater or lesser extent, and it feeds both the lungs and the body. Oxygen-rich blood returning from the lungs mingles to some extent with 'stale' blood returning from the rest of the body. Fishes have only a single circulation, with the blood going from the heart to the gills and then straight round the body before returning to the heart.

The rhythmic pumping of the heart, apparent in a person's pulse, goes on throughout life. The MUSCLES which make up the heart are of a very special kind, with branched and interlacing fibres which can go on contracting and relaxing without ever getting tired.

Heartsease, *see* PANSY

Heart urchin, *see* SEA URCHIN

Heartwood The central region of a tree-trunk, composed of xylem but with no living cells and not actually involved in the conduction of water — the vessels have been compressed and

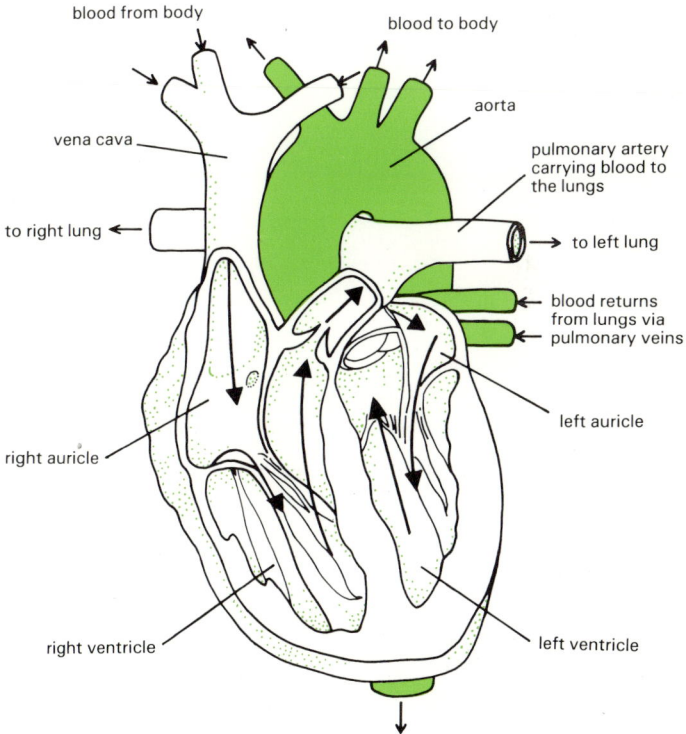

The human heart

Heath

impregnated with various resins that make the wood very hard and resistant to decay.

Heath (1) A region of acidic sandy soil with low fertility. There is usually a thin layer of dry peat and the dominant plants are members of the heather family. (2) A large group of low-growing evergreen shrubs, native to Eurasia and Africa. Heaths have needle-like leaves and small red or purple bell-shaped flowers. Some have white flowers. They cover open moorland and heathland.

Heather A low evergreen shrub of Eurasia and North America related to heath. True heather, also called ling, has purple flowers and small triangular leaves.

Hedgehog An insectivorous mammal living in Eurasia and Africa. The most obvious thing about a hedgehog is its prickly coat, composed of numerous hardened hairs. These spines are found only on the upper surface, the lower surface being clothed with perfectly ordinary soft hair. Hedgehogs feed on a wide variety of small animals and plant food. In cooler regions they hibernate for the winter.

Hedge sparrow, *see* ACCENTOR

Heliotrope A perennial shrubby plant, native to Peru. It has oval leaves and clusters of purple or lilac flowers.

Hellbender A giant salamander living in North America and reaching a length of up to 75 cm. It has a flattened, brownish body and a broad head with small eyes. The tail, which accounts for one-third of the animal's length, is oar-like at the end. The hellbender lives in swift streams in the eastern United States. Similar species live in China and Japan.

Heredity

TT

tt

parent plants

T   T   sex-cells   t   t

Tt

first generation

T   t   T   t

TT   Tt   Tt   tt

second generation

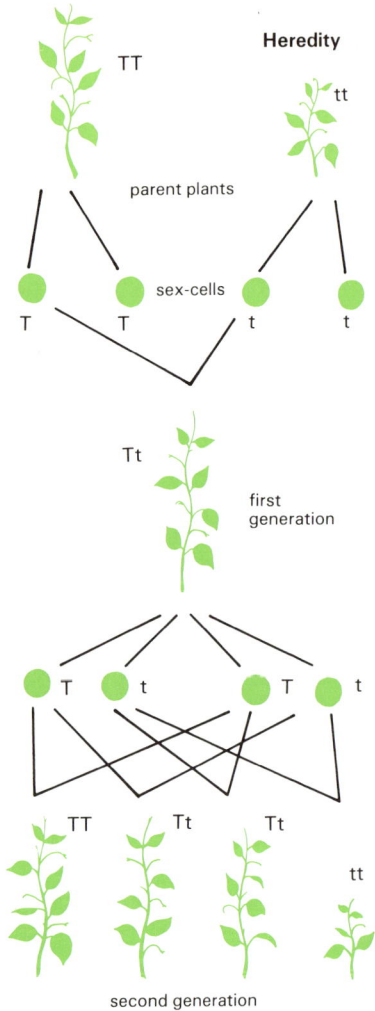

Hedgehog

Mendel found that when he crossed a tall and a short pea all the new plants were tall. But crossing these resulted in one short plant to every three tall ones. He explained this by saying that each plant carried two 'factors' for height and that the tall factor (T) was dominant over the short factor (t).

114

Hellebore Any of a group of perennial Eurasian herbs with large lobed leaves. Hellebores have green, white or purplish flowers, often with no petals, but with petal-like sepals instead. They are poisonous. The Christmas rose is a hellebore.

Hemlock A poisonous perennial herb of Europe and Asia. It has leaves like parsley and purple spots on its stems. It smells unpleasant. The flowers form white umbrella-like clusters. The name is also used for a group of small-coned coniferous trees scattered over the northern hemisphere and much grown for their timber.

Hemp A tall Asian plant, now grown also in Europe and North America, whose fibres are used to make rope and cloth. It has yellowish-green flowers and long, toothed leaves.

Henbane A poisonous herb with sticky, bitter leaves and yellow and purple flowers. It can be fatal to chickens and other animals that eat its leaves or seeds. The henbane is native to Eurasia and Africa.

Hepatic Concerning the liver.

Herb Any small plant with soft stems. Most herbs die back in winter and regrow from the roots in spring, but some are annuals. The term herb is also employed for certain edible plants, especially those which are used to flavour food.

Herbaceous Non-woody.

Herbivore An animal which eats mainly vegetable material.

Heredity The study of the ways in which the characteristics of a plant or animal are inherited, or passed on to its offspring. Gregor Mendel, a monk living in the middle of the 19th century, was the first person to study this subject systematically, and mendelism is another word for heredity. Mendel experimented with garden peas and grew thousands of plants in the monastery gardens at Brno, which is now in Czechoslovakia. After much painstaking work, Mendel concluded that characteristics such as the height of the plants and the colour of the seeds are controlled by minute particles which he called 'factors'. He thought that these 'factors' were carried to the next generation in the pollen and egg-cells.

Few people believed Mendel, however, and much of his work was destroyed after his death. Mendel had known nothing of GENES and CHROMOSOMES, and it was not until these structures were discovered long after Mendel's death that people realized he had been right. The genes are Mendel's 'factors'. Mendel's work is still the basis of plant and animal breeding — the rearing of bigger and better crops and animals.

Hermaphrodite Having both male and female parts in one individual. Many flowering plants are hermaphrodites as many animal groups, particularly earthworms and snails. Special features have been evolved which prevent self-fertilization.

Hermit crab A crab which lives in the empty shell of an animal such as a winkle or whelk. Hermits have a soft banana-shaped abdomen. Only the front end of the animal has a hard skeleton, and the right claw (larger than the left) is used to close the opening of the shell.

Heron A large wading bird found in most temperate and tropical parts of the world. The grey heron ranges from western Europe to South-East Asia. The great blue heron and the green heron are found over much of North and Central America.

Herring

Herring The herring has long been the most important food fish in the world. Several thousand million herrings are caught in the Atlantic and neighbouring seas every year. The fish is a greyish-green colour on top and ·silvery beneath. It swims in huge shoals near the surface of the sea.

Hessian-fly A small, gnat-like fly whose sap-sucking larvae are serious pests of cereal crops, especially wheat, in North America.

Heterodont Having teeth of various types (i.e. mammals) as opposed to the homodont condition of reptiles whose teeth differ only in size.

Hibernation In cold and temperate regions, many animals hibernate during the winter. This is a state of inactivity or deep sleep during which the body processes slow down almost to a stop. The body temperature, even in mammals, falls to within a degree or two of that of their immediate surroundings. Aquatic animals do not normally hibernate, although many become lethargic during the cold weather. Most soil-dwelling animals merely burrow deeper to avoid the cold, but many free-living invertebrates hide away for the winter. Probably the majority of insects pass the winter as eggs which are very resistant to drought and cold, but many over-winter as larvae, pupae, or adults.

Among vertebrates, amphibians and reptiles are well-known hibernators. Frogs, tortoises, snakes and lizards all bury themselves away from the effect of frost. True hibernation is not known among birds although the poorwill, an American nightjar, is known to become very drowsy and to sleep for long periods. Many mammals, too, hide away and remain drowsy during the winter months. Bears, badgers, tree squirrels and others go to sleep for varying periods of time, but they wake periodically and their temperature does not drop more than a few degrees below normal. True hibernation, where the body temperature falls almost to that of the surroundings, is found in only a few groups of mammals. The egg-laying monotremes and some of the opossums are known to hibernate in cold winters. Bats of temperate and cold climates hibernate because they cannot catch insects in winter. Some insect-eating mammals — notably the hedgehog — and many rodents (e.g. dormice, ground-squirrels and hamsters) also go into a deep winter sleep. Even so, these hibernators often wake up and may feed on stored food.

Hibiscus The name of about 200 plants of the mallow family. Most live in tropical regions. They have showy flowers, mostly bell-shaped with protruding stigmas and stamens, in colours ranging from red to white, with some yellow. American species are pollinated by humming birds. There are many cultivated varieties.

Hickory The name of 15 species of trees which grow in North America and eastern Asia. They have large compound leaves. Hickory trees have male and female flowers without petals and produce woody fruits. The seeds of some species, such as the pecan, are edible. Hickory wood is hard and tough and the trees are members of the walnut family.

Hinny A hybrid animal resembling a small horse which results from the mating of a stallion and a female ass.

Hippopotamus Distantly related to the pigs, the hippopotamus is one of the world's largest land animals after the elephants. It reaches a length of 4 metres and may weigh 4 tonnes. Hippos spend most of their time wallowing in rivers and lakes. They are found over much of the central part of Africa. The pigmy hippopotamus of West Africa is a separate species.

Histology The study of the structure and development of tissues.

Hive-bee, *see* HONEY-BEE

Hoatzin One of the strangest of all living birds, the hoatzin lives in the dense forests of the Amazon basin. It is related to the game birds and has brown plumage with a crest of long, spiky feathers. But the young hoatzin has claws on its wings and it uses these to climb about in the trees. It does not fly well for it has only small flight muscles.

Hobby A small falcon found particularly in scattered woodlands of the northern hemisphere where it chases small birds at high speed through the trees.

Hogweed The name of some large, coarse-leaved plants of the PARSLEY family. Hogweeds, sometimes called ragweeds, have hollow stems and clus-

The name hippopotamus means 'river horse'. Hippos spend much of their time in rivers, feeding and wallowing in the mud.

ters of small white flowers. They grow up to nearly 2 metres in height, and the giant hogweed reaches almost twice that.

Holly A hardy evergreen tree found in temperate parts of the world. It has glossy leaves, which are generally spiky, at least on the lower branches. The trees are either male or female: the female trees bear small red berries, though some are yellow. The timber is hard and close-grained.

Hollyhock A tall Chinese perennial herb, grown elsewhere for its spike of white, pink, yellow or purple flowers. It has large heart-shaped leaves. The plants grow up to 3 metres tall.

Homeostasis The maintenance of constant conditions within the body, e.g. constant temperature and constant composition of the blood.

Homoiothermic The condition found in birds and mammals where the body is kept at a constant temperature independent of that of the surroundings.

Homologous chromosomes Chromosomes which are structurally identical and which pair up during MEIOSIS. The genes present in these chromosomes affect the same body organs or developmental processes, but not necessarily in the same way. One gene, for example, may 'instruct' a plant to produce white flowers, while its homologous gene in the other chromosome may be for red flowers. In such instances, one gene generally dominates the other and suppresses its effect.

Homozygous The condition in which identical genes occur at the same point (locus) in each of a pair of HOMOLOGOUS CHROMOSOMES.

Honesty A biennial European herb with toothed heart-shaped leaves. The sprays of purple flowers mature to form broad silvery disc-shaped fruits. These discs contain the seeds.

Honey ant Found in the drier regions of the world, honey ants feed largely on the honeydew exuded by aphids. Much of the honey collected in time of plenty is stored in the bodies of special workers called repletes. These become greatly swollen and they hang from the roof of the nest chamber, giving out honey when stroked by other ants.

Honey badger, *see* RATEL

Hoopoe

Honey bear, *see* KINKAJOU

Honey-bee Also known as the hive-bee, this is the best known of all bees because it is bred and kept in domestic colonies. A bee's nest or hive may contain as many as 50,000 individuals of three different types — workers, drones and one queen bee. The queen bee does nothing but lay eggs. The drones are males who do no work; they are merely needed to mate with a new queen. The workers are females who are incapable of reproduction. It is they who build the hive, gather nectar and pollen and look after the other bees in the hive. Workers usually live only a few weeks. New queens result from grubs being fed on a protein-rich substance called royal jelly. The first queen to emerge from her cell kills the other queens while they are still in their cells and goes off on a 'marriage flight', during which she mates with a drone. She then normally returns to the hive and ousts the old queen who either leaves the hive with a swarm of workers to found a new colony or is killed by the workers.

Honey eater A large family of song-birds that feed mainly on nectar and insects. They are found in Australia, New Zealand and nearby Pacific islands.

Honeyguide The honeyguides are small greyish-brown birds living in Africa and southern Asia. They feed mainly on bees and wasps, and seem to be immune to the insects' stings. The birds get their name because some species actually attract honey badgers and lead them to the bees' nests. The badgers rip open the nests to get at the honey and the birds feed on the scraps, including the wax combs.

Honey-parrot, see LORIKEET

Honeysuckle Any of a group of climbing shrubs, found world-wide in temperate regions. It has dark green oval leaves and fragrant trumpet-shaped flowers. Most of the many species are evergreen.

Hooded seal This large seal gets its name from the inflatable bladder or hood behind the nose of the male, the purpose of which is unknown. Males grow to almost 3 metres and weigh up to 400 kg. Hooded seals are found from Newfoundland north to the Arctic Ocean.

Hookworm A small parasitic roundworm found in the tropics. It enters the body of humans and other mammals through the skin or via infected drinking water and lives in the small intestine, sucking blood.

Hoopoe A widely distributed Old World bird, the hoopoe has a distinctive fan-shaped crest and a long curved beak. Its name is derived from its call.

Hop A perennial herbaceous climber of the north temperate zone. It has heart-shaped lobed leaves with toothed edges. Male plants have small green flowers. The larger cone-shaped heads of the female flowers are used in making beer.

Hopper A large group of small sap-sucking insects. They include frog-hoppers (whose nymphs produce the bubbly froth known as cuckoo spit), leaf-hoppers, tree-hoppers and plant-hoppers. Many cause considerable damage to cultivated crops.

Hormone A complicated substance produced by certain cells (normally collected into glands) that is released into the bloodstream continuously or periodically. Many hormones are known — e.g. adrenaline, insulin, thyroxine — each with a specific composition and activity. Hormones are extremely important in the regulation of the internal activity of the body.

Hornbeam Any of several species of large deciduous trees of northern temperate forests related to the BIRCH. It grows up to 18 metres tall, has small oval leaves and produces catkins. The wood is very hard. The hornbeam is commonly planted for hedges.

Hornbill Hornbills are large birds found in Africa and South-East Asia. They have huge bills, often topped by large bony swellings called casques. Most of the species live in the forests and feed on fruit. They nest in hollow trees and, after laying her eggs, the female shuts herself in with a wall of mud. Only a small slit is left, and the male feeds his mate through this.

Horned toad A small lizard of North America with a toad-like face and thorn-like scales on its body.

Hornet This name is commonly applied to almost any large wasp, but the true hornet is a particular species living in wooded areas of Europe. It differs from the ordinary wasps in being brown and yellow instead of black and yellow, but it leads just the same kind of social life. It builds its nests in hollow trees. Hornets have a powerful sting, but are less ready to use it than the wasps.

Horse Wild horses were widespread over Europe and Asia in prehistoric times. The only survivor is Przewalski's horse of central Asia, a stocky creature 1.4 metres high at the shoul-

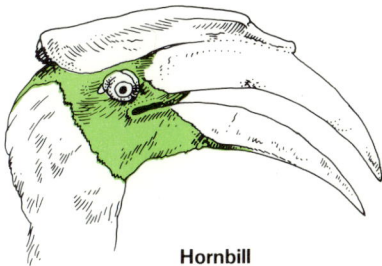

Hornbill

der. The horse was domesticated some 4000 years ago and there are now more than 60 distinct breeds.

Horse chestnut Any of a group of trees and shrubs of Asia, Europe and North America, where it is called buck-eyes. It is not related to the sweet CHESTNUT. The large compound leaves of the horse chestnut are fan-shaped and it has candle-like spikes of white or red flowers. The seeds, borne in spiny fruits, are the conkers of schoolboys and are not edible by humans. The wood is soft.

Horse-fly The name given to certain large flies whose females feed on the blood of animals, especially cattle and horses. The males suck nectar from flowers. Cleg-flies are related to the horse-flies.

Horse mushroom, *see* MUSHROOM

Horseradish A member of the cabbage family, horseradish is a perennial European herb with large lance-shaped leaves and very deep tap roots. The white flowers are borne on stems 60 cm tall. The roots are ground and used to make a spicy sauce.

Horseshoe bat Many bats have folds of skin on their faces called nose-leaves. The horseshoe bat gets its name from the fact that one part of its nose-leaf is shaped like a horseshoe. There are more than 50 species in temperate and tropical parts of the Old World. The largest has a wingspan of almost 40 cm.

Horseshoe crab, *see* KING CRAB

Horsetail A group of non-flowering plants related to the ferns. Only about 25 living species are known in the world today and most of them are small plants about 30–45 cm high. Horsetails were very common millions of years ago and the remains of many large, woody forms are often found in coal. The horsetail plant consists of a branching, underground rhizome and a number of upright aerial shoots with or without whorls of slender branches. The rhizome lasts from year to year but its roots and the aerial stems are usually renewed each year. The horsetail reproduces by scattering spores from simple cones at the tips of the aerial stems. It has no true leaves.

Host An organism carrying a parasite.

House-fly A familiar insect found in and around houses world-wide. It feeds on any organic material, dissolving it if necessary with saliva and sucking up the solution with its short proboscis. Houseflies spread disease since they are equally happy feeding on refuse as on edible food. The larvae develop in dung and other rotting matter.

Houseleek A perennial succulent herb growing in the mountainous regions of Europe and western Asia. It has a rosette of flat, fleshy leaves. The flowering shoots are up to 60 cm tall, with a cluster of pink or yellow blooms. People used to plant it on rooftops as a protection against lightning. There are several species.

House mouse Probably the most common and widely distributed rodent of all, it is found wherever people have built houses. Despite their bright, beady eyes these mice are short-

Horsetail

sighted, and rely far more on their senses of hearing and smell. The house mouse is an adaptable feeder, but its success lies also in the high breeding rate which averages five litters a year with five young.

Hover-fly A group of true flies which can move in all directions as well as hover. Many of them mimic bees and wasps, but the hover-flies can always be distinguished by their single pair of wings. Each wing has a 'false margin' formed by a vein running near the edge.

Howler monkey The largest of the South American monkeys, the howlers are named from their loud calls which are produced in a large 'echo chamber' in the throat. The monkeys are up to one metre long, including the tail, and range from red to black in colour.

Humerus The bone of the upper fore-limb, between the shoulder and elbow in man.

Humming bird There are over 300 species of these little birds, distributed throughout the New World. The largest is a mere 20 cm long, and the smallest has a body no bigger than a bumble-bee. Most of them have brilliant plumage. They get their name from the noise made by the wings as the birds hover in front of flowers. Most of them feed on nectar, sucking it up through the tubular tongue and narrow pointed beak.

Humpback whale Distinguished by unusually long flippers, this baleen whale grows to 15 metres in length. It makes regular migrations between tropical waters in the winter and polar waters in the summer and can frequently be seen in coastal waters.

Humus The decaying remains of animals and plants in the soil (soil organic matter). The soil bacteria are constantly changing the humus. As it decays different substances are formed. Plants will be able to absorb some of these (e.g. nitrates). Humus is essential to productive soil. Besides supplying minerals it conserves moisture and loosens soil particles to admit air and water.

Hyacinth Bulbous flowering plants of the lily family, originally from the Mediterranean region. Each bulb has

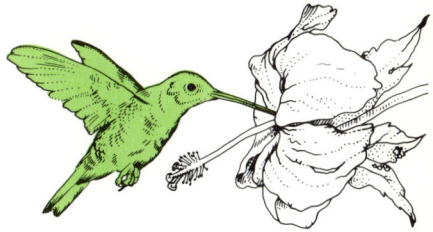

Humming bird

long fleshy leaves and a spike of bell-shaped, sweet smelling-flowers in many colours.

Hybrid An organism resulting from the mating of two different species. Only closely related species can mate and produce offspring and even then the offspring itself is normally sterile. The mule, offspring of a horse and an ass, is one of the best known animal hybrids. Many cultivated flowers are hybrids: they are more likely to be fertile than animal hybrids.

Hydra A tiny freshwater animal related to the sea anemones. It has a tubular body ending in a ring of tentacles containing stinging cells which are used to capture prey such as water fleas.

Hydrangea Any of a group of flowering shrubs of the Americas and eastern Asia. Hydrangeas have large, long leaves and produce heads of showy white, pink or blue flowers.

Hydroid, *see* SEA FIR

Hydroponics The science of growing plants in nutrient solutions without soil.

Hydrozoan A member of the coelenterate class Hydrozoa. These animals are usually colonial and most show a well-marked ALTERNATION OF GENERATIONS. The PORTUGUESE-MAN-OF-WAR belongs to this class and each 'individual' is, in fact, a floating colony composed of numerous polyps. HYDRA is also a member.

Hyena There are three species of hyena: the spotted or laughing hyena, the brown hyena, and the striped hyena. The latter extends from North Africa to India, but the other two are found only in the southern half of

Africa. They are greyish-brown creatures, up to 90 cm high at the shoulder. The back slopes sharply down to the hindquarters. The head is large and the jaws are very powerful. Hyenas hunt in packs and kill many antelopes. They also eat lots of carrion.

Hypha (plural hyphae) A filament or thread of a fungus.

Hypogeal That type of germination in which the seed leaves stay below ground (e.g. runner bean).

Hypogynous flowers are those in which the petals arise below the carpels.

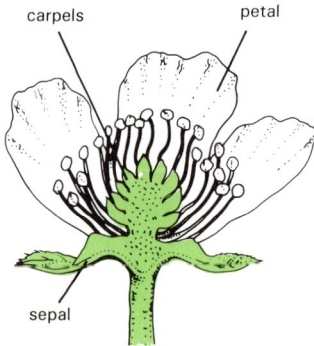

Hypogynous flower

Hyrax The hyraxes or conies are so unlike any other animals that they are given a whole order to themselves. They are greyish-brown tailless mammals with short muzzles and small rounded ears. The largest species is only 45 cm long. It is thought that the elephants are probably the nearest relatives of the hyraxes. Hyraxes live in Africa and south-western Asia and feed on grasses and other low-growing plants.

Hyssop An aromatic bushy evergreen shrub native to the Old World. It has a square stem, small lance-shaped leaves, and spikes of bluish flowers. It was formerly used as seasoning and as a medicine.

I

Ibex The name given to several species of mountain-dwelling goats. They range from Spain to Mongolia. Ibexes are sturdy animals, usually with long backward-curving horns. The horns, which are shorter in the females, generally carry prominent knobs on the front edge.

Ibis A widely distributed group of birds related to the herons and flamingoes. They all have long legs, fairly long necks, and long curved beaks. Many are brightly coloured, such as the scarlet ibis of tropical America. Ibises live in marshes and around lakes, feeding on aquatic animals. They nest in trees, however, just like the herons.

Ichneumon The name given to a number of wasp relatives which lay their eggs on or in the larvae of other insects. When the eggs hatch, the ichneumon larvae feed on the body of their host.

Iguana A group of large lizards found in tropical America and on the Galapagos Islands. They range in size from 1.2 to 1.5 metres including the long tail. Some of the species are caught for food.

Ibex

Imago The final stage of an insect's life-history — a mature adult.

Immunity The ability to resist attack by, or the effect of, a parasite.

Impala A graceful antelope about 90 cm high at the shoulder. It is chestnut brown, with a sharply defined white belly. The male has horns up to 75 cm long, but the female is hornless. The impala inhabits a large part of eastern and southern Africa, usually keeping near to water.

Incisor One of the chisel-shaped cutting teeth at the front of the mammalian jaw.

Indehiscent Fruits which do not split to release seeds (e.g. hazel-nut). They generally contain just one seed, which breaks open the fruit as it starts to germinate.

Indian antelope, *see* BLACKBUCK

Indian bear, *see* SLOTH BEAR

Indian buffalo Also known as the water buffalo, the Indian buffalo is an aggressive animal in the wild. But the domesticated buffalo is an important draught animal in much of Asia. It is a docile but powerful creature weighing up to one tonne.

Indian hill mynah, *see* MYNAH

Indigenous Native to a given area — not introduced.

Indri The largest of the lemurs, the indri reaches 75 cm in length. It has very long hindlimbs, but its tail is little more than 2 cm long. The fur is basically black and white. The indri lives only in the northern parts of Malagasy (Madagascar), where it dwells in the trees and feeds on leaves.

Inflorescence The whole flowering part of a plant — including its stalk — is called an inflorescence. Inflorescences are of varying kinds, some of which are illustrated. A solitary flower is one of the simplest, while a composite flower is made up of many florets — tiny flowers clumped tightly together.

Inkcap The name of a group of toadstools with conical or bell-shaped caps. When ripe they disintegrate into an inky fluid.

Insect The arthropod class Insecta is the largest of all animal groups, with something like a million known species and many more certainly yet to be dis-

covered. Yet even this vast number can be reduced to a fundamental pattern.

The body has three regions: head, thorax and abdomen. The head bears one pair of antennae, a pair of compound eyes and a number of mouthparts which assist in feeding. Behind the head, the thorax bears three pairs of legs and usually two pairs of wings. The abdomen has no limbs. All adult insects breathe air and have numerous fine tubes (tracheae) branching throughout the body. The tracheae carry air to the tissues from the spiracles along the sides of the body. Insects are limited in size by the

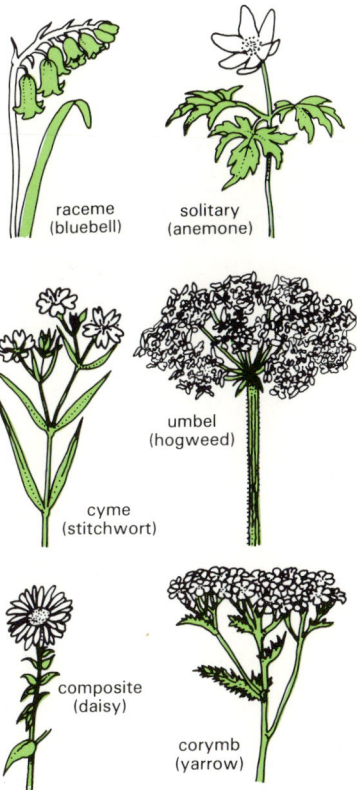

raceme
(bluebell)

solitary
(anemone)

umbel
(hogweed)

cyme
(stitchwort)

composite
(daisy)

corymb
(yarrow)

Inflorescences

The parts of a typical insect

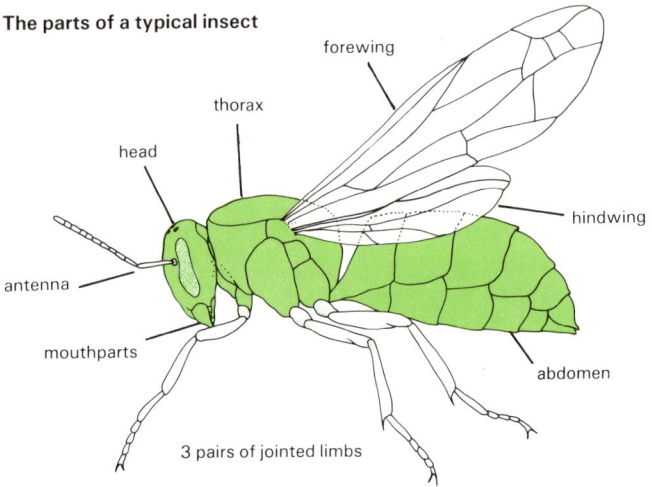

forewing

thorax

head

hindwing

antenna

mouthparts

abdomen

3 pairs of jointed limbs

tracheal system, as air cannot diffuse efficiently for more than a short distance along these tubes. Insect bodies are therefore never very thick. The African Goliath beetle is one of the largest insects, with a body about 10 cm long and 5 cm thick. Butterflies and moths with large wing spans have relatively narrow bodies.

Most insects lay eggs, but a few give birth to active young. Because of its hard cuticle, an insect cannot grow steadily as a vertebrate does. Periodically the insect gets rid of the old 'skin' and replaces it with a new larger one. The new cuticle is formed under the old and then by swallowing air or water, the insect bursts the old one. The new cuticle soon hardens on contact with the air and then, by getting rid of the excess air or water, the insect leaves enough room for the new growth period. The process of changing the skin (moulting) is called ecdysis. It normally occurs between three and eight times during the insect's life. Mayflies with a long growing stage may moult up to twenty times. When once the adult stage is reached, no further growth occurs. Frequently the young insect is very different from the adult. (Compare a

caterpillar and an adult butterfly.) The young insect is, in this case, called a larva. Before it can assume the adult form it must go through a period of rest during which the necessary changes can occur. This period is the pupal period. The change from larval to adult form is called metamorphosis. When the young insect is just a small version of the adult (e.g. a grass-hopper) it develops gradually into an adult. This young form is called a nymph and its metamorphosis is called partial or incomplete, as opposed to the complete metamorphosis of a butterfly. The life-cycle from egg to adult may be very short — a housefly can, in warm conditions, develop from egg to adult in little more than a week. On the other hand, mayfly nymphs live for two years or more before the adult emerges for its brief life in the air. Some wood-boring moth larvae also live for several years before reaching maturity. Insects are not normally active in the colder months of the year. They may overwinter in any of the life-stages.

Insects are very successful animals, having colonized land, water and air. Only the sea, with very few insect inhabitants, has proved a serious barrier to their spread. The main factor in their success on land has been the waterproof cuticle that allows them to inhabit dry places. And their small size enables them to exploit numerous habitats which are unsuitable for larger animals. These advantages, combined with the power of flight, have helped to establish insects firmly all over the Earth.

Insectivore A member of the mammalian order Insectivora which includes hedgehogs, moles and shrews. These animals do not feed exclusively on insects — worms, slugs, and even some vegetable matter form part of an insectivore's diet.

Instar Any stage in an insect lifehistory between two moults.

Instinct An inborn pattern of activity found in some form or other in almost all animals. Instincts account for much of animal behaviour — courtship displays, protective care of the young, migratory drives and reaction to dangers. Instincts like nest-building and web- and cocoon-spinning are perfect from the start. Caterpillars of different species of moth spin their own types of cocoon once only in their lives — but they do it perfectly. Young birds reared in isolation away from parents nevertheless build exactly similar nests even down to the material used. Other instincts, like man's instinct to walk, take time and practice to become perfect. A bird's song also needs practice and is partly learned from its parents: in other words, a bird instinctively knows how to sing but has to learn the *right* song. All members of a species will usually respond in much the same way to a stimulus. Instincts are just as much a part of an animal as the structures which identify its body.

Insulin A hormone secreted by the 'Islets of Langerhans' in the pancreas of mammals and other vertebrates and involved in controlling the amount of glucose sugar in blood. When the sugar level is high — after a meal, for example, the insulin causes the liver to store the excess glucose as glycogen (animal starch). Another hormone, called glucogon, causes the liver to release glucose again when the blood sugar level falls.

Intestine That part of the alimentary canal beyond the stomach which is concerned with the later stages of digestion and with the absorption of food into the bloodstream. It also reabsorbs water from the faeces.

Invertebrate An animal without a backbone. The majority of animals are invertebrates, the backboned, or vertebrate animals being fishes, amphibians, reptiles, birds and mammals.

Iris About 150 species of perennial herbs growing from rhizomes, corms or bulbs. Those growing from rhizomes are often called flags. Irises have fleshy, lance-shaped leaves and large showy flowers in all colours.

Iris

Irish moss, *see* CARRAGHEEN

Ironbark The name given to most species of eucalyptus and also a West African tree. The West African ironbark has leathery leaves, white flowers and very sweet berries called 'Miraculous berries'.

Ironwood A name for many kinds of trees with very hard timber, especially American HORNBEAM.

Irritability A property of all living things — the ability to react to changes in the surroundings.

Ivy A climbing shrub of Europe, North Africa, Asia and North America. It has thick, glossy leaves, often with five lobes, yellowish-green flowers and small black berries. It climbs by means of clusters of tiny roots that grow from the stems and take hold in any small crevice.

J K

Jacamar A family of small birds living in the forests of Central and South America. Most have long bills, long tails and iridescent plumage. They feed on insects and dart from their perch to catch their prey on the wing.

Jacana Also called the lily trotter, the jacana is a water bird related to the curlew, although it looks more like a long-legged coot. It walks over the floating leaves, supported by its long spreading toes. Jacanas are found throughout the warmer parts of the world.

Jacaranda Any of a group of trees and shrubs of tropical America. They are part of a larger group known as bignonias. They have large showy flowers, and are widely· cultivated in many warmer parts of the world.

Jackal A fox-like member of the dog family. Two species live in Africa, and a third lives in North Africa, Asia and south-eastern Europe. Jackals range from black to dirty yellow in colour and stand about 40 cm at the shoulder. They feed on the left-overs from lion kills, but also hunt gazelles and other small animals.

Jackass hare, *see* JACK RABBIT

Jackdaw A member of the crow family, the jackdaw is well-known for its habit of stealing bright objects. It is about 33 cm long and has black plumage shot with blue on the back and head and a grey nape. It has a short strong bill and feeds on insects, small vertebrates, and fruits and seeds. It ranges across Europe and Central Asia.

Jack Dempsey A small freshwater fish of South America. It became popular with aquarists at the time Jack Dempsey was the world heavyweight boxing champion and was named after him because of its aggressiveness.

Jack-knife clam, *see* RAZOR SHELL

Jack rabbit The name given in America to various hares of the western United States including the plains or prairie hare and the jackass hare.

Jaguar The largest of the American cats, the jaguar is about the same size as a leopard but more heavily built, weighing up to 150 kg. The coat is yellow with black spots arranged in rosettes of 4–5 around a central spot. It lives mainly in the dense forests of South America.

Jaguarundi Although a member of the cat family, the jaguarundi is more like a large weasel in shape and habits. Its body is long, up to 1.25 metres, but with a shoulder height of only 30 cm. It has short, rusty-red or grey fur. The jaguarundi lives on the edge of forests from Mexico to Argentina.

Japonica A name used for any plant first discovered in Japan, but especially for three species of Japanese QUINCE. They are hardy deciduous shrubs, with toothed leaves, producing edible fruits.

Jasmine Any of 200 species of shrubs or climbers, related to the OLIVE. They originated in warmer parts of the Old World. Many are grown for their star-like flowers. An oil used in perfume is extracted from common jasmine.

Jay

Jay Colourful members of the crow family, found in North America, Europe and Asia. The European jay has reddish-brown plumage, darker on the back, with black and blue-barred wings. It lives in woods and forests but often feeds in open scrubland. The blue jay is one of the best known American species.

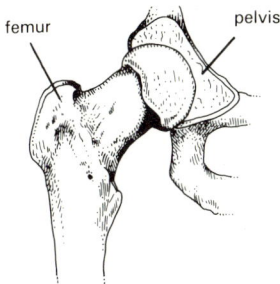

Ball and socket — hip joint

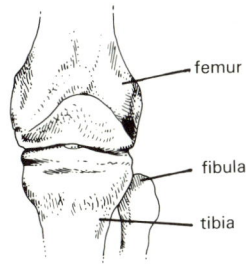

Hinge — knee joint

Jellyfish Marine animals with a transparent, umbrella-shaped body fringed with tentacles. Most have stinging cells and capture fishes and other animals. Some merely suck in small planktonic creatures. The largest, found in Arctic waters, has a body up to 2 metres across, but the most venomous jellyfishes are the small 'sea wasps' of the western Pacific.

Jerboa A group of small desert rodents found in Africa and Asia. All have long hindlegs and hop like kangaroos.

Jew's ear fungus A brown, ear-shaped fungus found growing on elder trees. It is covered with fine grey hairs. In China it is esteemed as a food.

John Dory A marine fish whose body is flattened from side to side into a disc-like shape. On each flank is a distinctive black spot edged with yellow. The John Dory lives in the Mediterranean and eastern Atlantic.

Joint There are several types of joint in the vertebrate body, especially in the body of a mammal. One type is called a hinge joint. The two bones meet to form a hinge and can move in one plane only. Elbows and knees are hinge joints. Another is the ball and socket joint. This is found in hips and shoulders. The rounded ends of the leg and arm bones fit into cup-shaped hollows, allowing a wide range of movement.

Judas tree A deciduous tree of southern Europe and western Asia. It grows to a height of 9 metres, and has clusters of bright pink flowers which usually open before the smooth heart-shaped leaves. Its name comes from the legend that it was on one of these trees that Judas Iscariot hanged himself.

Jumping bean The fruit of members of the spurge family, growing in Central and South America. They jump about because of the movements of caterpillars living inside them.

Jungle fowl A group of forest-dwelling birds belonging to the pheasant family which live in India and South-East Asia. They include the red jungle fowl from which the domestic chicken has been bred.

Juniper Any of 60 species of trees or shrubs of the CYPRESS family. Junipers are evergreens, with reddish-brown bark and sharply pointed leaves. They produce berry-like, brown or greyish-blue cones which are used to flavour gin.

Jute The name of two species of annual herbs grown in India and Pakistan. They have cylindrical stalks up to 4.5 metres tall, with tapering leaves and small yellow flowers. From the stalks come strong fibres used to make coarse fabric, string and rope.

Kagu Found only in the forested highlands of New Caledonia, the kagu is a bird with no close relatives. It is somewhat larger than a domestic fowl and has pale grey plumage. Like the domestic fowl, the kagu seldom flies, running fast with wings outstretched when alarmed. Kagus are becoming increasingly rare.

Kestrel

Kakapo An extremely rare, owl-like parrot living in the rain forests in parts of New Zealand. It is a flightless bird about 50 cm long with greenish upper parts streaked with black, brown and grey.

Kale A form of cabbage with loose, curly leaves. It is cultivated for human and animal food. Marrow-stem kale has particularly thick stems and is grown for feeding sheep. Sea kale is a totally different plant.

Kangaroo A group of Australian marsupials with powerful hindlimbs, short forelimbs and a strong tail. Best known are the great grey kangaroo of eastern Australia and the red kangaroo found in most parts of the continent. The former can measure 2 metres in height and is a browsing animal of the open forests. The red kangaroo is similar in size and grazes on the open plains. Kangaroos can hop at speeds up to 40 km/h and have been known to leap 8 metres in a single bound.

Kangaroo rat This is a rodent, named for its long hindlegs and tail, short forelegs and leaping gait. There are several species in North America, ranging up to about 50 cm in length, including the long tufted tail.

Kapok A tall evergreen tree of the West Indies, cultivated there and in southern Asia for its soft cotton-like fibre. The tree has a prickly stem up to 30 metres tall and leaves divided into lance-shaped leaflets. It has yellow flowers. The fibres are the hairs which clothe the seeds inside the fruits.

Katydid This is the name given to various American bush crickets. The males make remarkably loud sounds by rubbing their wings together.

Kauri pine A large coniferous tree of Australia and New Zealand. It grows up to 45 metres tall, and is much used for its timber and resin. The bark is pale and flaking.

Kea An olive-green parrot with orange markings on the underparts found in South Island, New Zealand.

Kelp A name often used for any brown seaweed but generally for very large brown seaweeds growing off the shores of Brittany, Ireland and Scotland, and the Pacific coasts of America. They are a source of iodine. The longest kelps are certain Pacific species which can exceed 50 metres.

Kestrel A small falcon noted for its hovering flight. There are several species, distributed over much of the world. All are brown and grey, with black spots. The common kestrel lives in Eurasia and Africa. It nests and roosts in open woodland, but hunts over open country. It hovers in the air and swoops on its prey.

Kidney The organ in vertebrates used to excrete waste nitrogen-containing substances produced by the breakdown of proteins in the body. The main waste material in man and other mammals is urea. This is filtered from the blood as it passes through the kidneys and passed out of the body in the urine. The kidneys also regulate the water content of the blood.

Killer whale A large black and white dolphin, up to 7 metres in length, which roams the oceans, particularly the colder waters. Killer whales feed on

Kingcrab

Cutaway of the human kidney

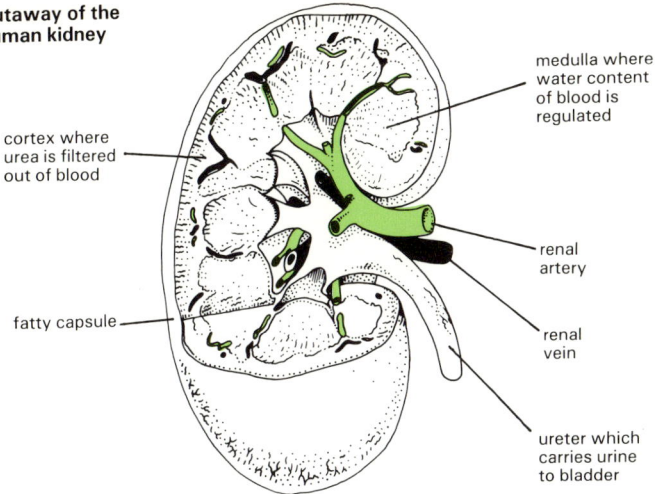

cortex where urea is filtered out of blood

medulla where water content of blood is regulated

renal artery

fatty capsule

renal vein

ureter which carries urine to bladder

other whales, seals and (and in the Antarctic) penguins. Despite their apparent ferocity, killer whales have been hand-tamed in oceanaria.

King crab This marine creature is not a crab, but a 'living fossil' more closely related to the spiders. It is also called the horseshoe crab because of its horseshoe-shaped body. The front end is covered with a horny shield, and the hind end bears a long movable spike. The whole animal may be up to 60 cm long. King crabs are found off the coasts of North America and eastern Asia.

Kingfisher A widespread group of stocky birds with long beaks and short tails, and often with brilliant plumage. Not all of them actually catch fish. Many of the species sit on perches and dart after lizards and insects on the ground.

King herring, *see* CHIMAERA

King penguin Closely resembling and closely related to the EMPEROR PENGUIN, the king penguin is slightly smaller and lives in ice-free seas of the sub-Antarctic.

King snake A group of non-venomous North American snakes up to 2 metres in length with a wide variety of colours and markings. They suffocate and eat smaller animals. Their prey includes other snakes such as the rattlesnake to whose venom they are immune.

Kinkajou A relative of the panda, the kinkajou or honey bear ranges from South America to Mexico. It has a body about 30 cm long and a prehensile tail half as long again. It lives in the treetops and uses the tail for extra support. The kinkajou has soft woolly fur coloured dark gold to brown. It feeds mainly on fruit.

Kingfisher

Kissing gourami Popular aquarium fishes from South-East Asia which have the habit of placing their lips together as though 'kissing'.

Kite A bird of prey with long pointed wings and a forked tail. There are three groups of kites which occur in Europe, North America, Asia and Africa. Kites live mainly in wooded valleys and they feed on fish as well as on many other animals and on carrion.

Kittiwake A small gull which nests in large colonies on cliff ledges and ranges the open seas of the North Atlantic.

Kiwi Small flightless birds of New Zealand. They are about the size of a chicken, but they have long curved beaks with the nostrils at the tip. Kiwis live in the forests and come out at night to search for worms and insects. The eyes are small, and the ears and nose are the main sense organs.

Kiwi fruit A climbing plant native to China, now cultivated in New Zealand; it is also called Chinese gooseberry. It has dark green heart-shaped leaves and creamy-white flowers. The hairy brown fruits are edible.

Klipspringer A small antelope of eastern and southern Africa which is renowned for its ability to scale seemingly impossible rock faces. The name means 'rock jumper' in Afrikaans.

Knapweed A European perennial herb, like a thistle without prickles. It has wiry stems, deep purple flower heads, and small leaves. There are several species, some of which are very large.

Knife-fish The name given to a number of unrelated freshwater fishes from South America, Africa and tropical Asia. All have flattened, blade-like bodies with a pointed hind end.

Knotweed A plant native to Asia with a cane-like stem and large heart-shaped leaves. It has sprays of small white blooms. Knotweed was introduced into Britain as a garden plant but has since become a widespread weed.

Koala

Koala The koala, one of Australia's most famous animals, is a pouched mammal related to the opossums. It looks like a small grey bear, with a prominent beak-like snout and tufted ears. Koalas are nocturnal creatures and feed on the leaves of eucalyptus trees. They rarely come down to the ground. They live mainly in eastern Australia.

Kob An antelope of the African savanna with lyre-shaped horns. A male kob stands just under one metre at the shoulder and weighs 90 kg. The female is slightly smaller.

Kodiak bear, see BROWN BEAR

Kohlrabi Derived from wild cabbage, the kohlrabi has a turnip-like edible stem.

Kola A tropical African tree with large, leathery evergreen leaves. It grows up to 12 metres tall, and has sprays of pale

Kiwi

yellow flowers. Its fruit contains fleshy red or white seeds (cola nuts) which contain caffeine, and were used to make stimulating drinks.

Komodo dragon, *see* MONITOR LIZARD

Kookaburra A drab member of the kingfisher family, also known as the laughing jackass because of its screams and chuckles. It is found throughout eastern Australia and has been introduced to the south-western region.

Krill This is a Norwegian word, used to describe the food of the whalebone whales. It refers especially to the little shrimp-like creatures called euphausids that float in millions in the colder seas of the world. They are about 5 cm long, but they feed the world's largest animal — the blue whale.

Kudu The kudu are among the largest of the antelopes. There are two species — the greater and the lesser kudu — and both have long twisted horns in the male. The greater kudu stands up to 1.5 metres at the shoulder, while the lesser kudu is up to one metre. Both have white stripes on the flank. The greater kudu ranges from the Sudan to South Africa; the lesser kudu is found only in East Africa.

Kumquat A dwarf evergreen tree, native to China. It is related to the ORANGE, and its fruits look like tiny round or oval oranges. The tree grows about 3 metres tall.

L

Labium The 'lower lip' of insects, actually composed of a pair of appendages fused together. This origin is fairly obvious in primitive biting insects such as the cockroach but the labium is sometimes extremely modified, as in the bugs where it forms the sheath of the 'beak'.

Labrum The 'upper lip' of insects, formed not from the paired appendages but from a part of the head above the mouth. Like the labium, it is sometimes extremely modified.

Laburnum A small deciduous tree native to southern Europe. It has glossy green leaves and long bunches of yellow pea-like flowers. The seeds and other parts are very poisonous.

Lacewing A group of delicate insects with two pairs of thin, netted wings. The best known ones are the green lacewings, or golden-eyes, but there are also many brown species. Both adult and larva feed on aphids.

Lacewing

Lachrymal gland A tear-producing gland found in all land vertebrates. In human beings, it lies beneath the upper eyelid. The tears are small quantities of antiseptic water produced to keep the surface of the eye moist. The lachrymal gland may be strongly stimulated in man by extreme emotional disturbance — the process of crying.

Lacquer tree Also called varnish tree, the lacquer tree is a small species of SUMACH native to China. It has long oval leaves and clusters of small flowers. It produces a resinous milky sap which is tapped and used as a varnish.

Ladybird A family of small, brightly coloured beetles, most of which are oval or circular in outline. They nearly all have red and black or yellow and black patterns. The adults and larva both consume greenfly and related pests.

Lamarckism An evolutionary theory (now disproved) put forward by the French scientist Lamarck (1744–1829). According to this theory, if a man trained hard for athletics and built up powerful muscles, his offspring would also have powerful muscles. In other words, Lamarck suggested that characters acquired during a lifetime could be inherited. It is certainly true that constant use of muscles develops them and that unused muscles deteriorate, but there is absolutely no evidence that these features can be inherited.

Lamina The blade of a leaf.

Lammergeier A graceful vulture found from the Mediterranean region to South Africa and China. It has a wingspan of up to 2.75 metres and, unlike other vultures, it has feathers on its neck. It is greyish-black on the back and rusty-brown underneath. The tail is long and diamond-shaped.

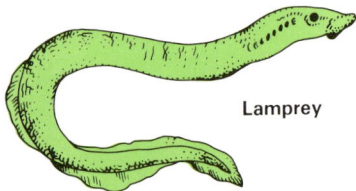

Lamprey

Lamprey A jawless, eel-like creature found in fresh and salt water in the temperate parts of the world. Young lampreys, and some adults, feed by sucking up debris from the sea- or river-bed. Others are parasites and feed by rasping flesh from other fishes.

Lamp-shell, see BRACHIOPOD

Lancelet A small marine creature about 5 cm long. It looks like a fish, but it has no paired fins and it is not a back-boned animal. It has some similarities with the vertebrates, however, and is probably close to the origin of back-boned animals. It feeds by straining food particles from the water.

Lanceolate (of a leaf) Elongated like a willow leaf but not linear as a grass leaf.

Langur A group of brightly-coloured monkeys with crests of hair on the head and patches of naked skin. Most are slender animals, about 60 cm long and sporting a tail of about the same length. They live in the forests of South-East Asia and, with one exception, they all keep to the trees.

Lapwing A large plover with a distinctive crest on the back of its head and black and white plumage. Also

Lapwing

known as the peewit from its call, the lapwing is a familiar bird of fields and open ground in much of Europe and Asia.

Larch Any of 12 trees related to the PINES. Unlike other coniferous trees, larches are deciduous. They grow up to 30 metres tall and bear their leaves in small whorls on short spurs.

Lark A group of small, dull-coloured, ground-nesting birds. One of the best known is the skylark which soars straight up from the ground and sings while hovering high in the sky.

female cones

Larch

sagittate

pinnate

orbicular

runcinate

lanceolate

cordate

perfoliate

oblong

ovate

connate

bipinnate

palmate

peltate

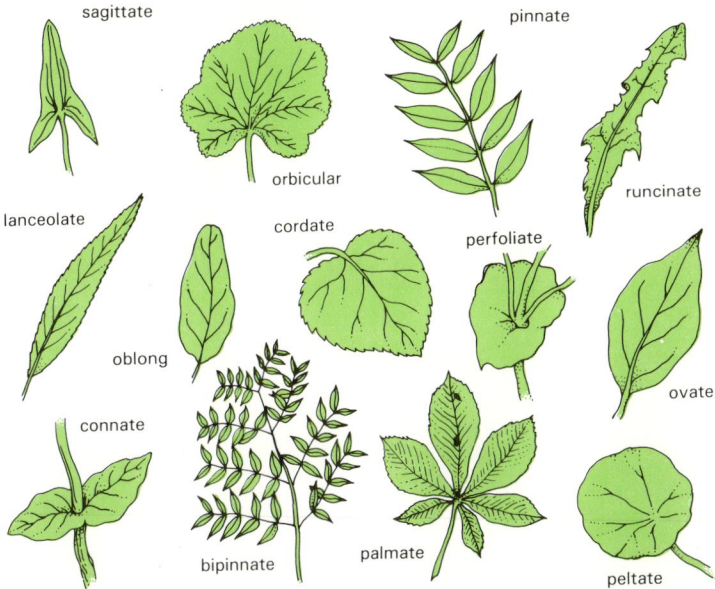

A variety of leaf shapes

Larva The young form of an animal that is very different from the adult. Examples include caterpillars, tadpoles and the various crustacean larvae. Some animals pass through more than one larval form before reaching maturity.

Lateral line, *see* FISH

Latex The white milky secretion of some plants such as rubber and dandelion.

Laughing jackass, *see* KOOKABURRA

Laurel The name given to a great many evergreen trees and shrubs of various families. The true laurel is the BAY-TREE, but the name is most often applied to the cherry laurel, a bush or small tree with shiny evergreen leaves.

Lavender A small fragrant bush of the mint family, native to Mediterranean countries. It grows up to 1.2 metres tall. It has slender grey-green leaves and long spikes of mauve flowers. It is used to make perfume.

Laver, *see* RED ALGAE

Layering, *see* PROPAGATION

Leaf Leaves are the plant's 'food factories' in which the vital process of photosynthesis is carried out. A typical leaf consists of a stalk called the petiole and a flat blade called the lamina. The latter bears numerous breathing pores (STOMATA), mainly on the lower surface. Leaves grow from the nodes of the stem and are arranged in such a way that no leaf is completely overshadowed by another on the same stem. In this way they intercept the maximum amount of light, which is necessary for photosynthesis. Leaves vary tremendously in shape, some of the different forms being shown in the diagram.

Leaf beetle A large group of small, often brightly coloured beetles which feed on the leaves of plants. Many are crop pests, one of the most notorious being the COLORADO BEETLE, a scourge of potato crops in some parts of the world.

Leafcutter ant A group of ants remarkable for the way they cut pieces out of leaves and carry them back to their underground nests. There the leaves are used to make a compost in which a fungus is grown. The ants feed solely on this fungus. Leafcutter ants are found in the warmer parts of the New World.

Leaf-hopper, *see* HOPPER

Leaf insect A group of insects with flattened bodies and leaf-like flaps on their limbs. These insects look just like leaves and are well camouflaged among the foliage in which they live. They belong to the same order as the stick insects (Phasmida) and are found in tropical regions.

Leatherjacket, *see* DADDY-LONG-LEGS

Leathery turtle This is the largest of the sea turtles. The foreflippers are also very large and a leathery turtle 2 metres long may have flippers spanning 2.75 metres. Unlike other turtles, the shell is made up of hundreds of bony plates covered with a leathery skin. The animals are found mostly in the warmer seas of the world.

Leech Related to the earthworms, leeches have a strong sucker at each end of the body. Most of them live in water and feed on fishes and snails.

Some suck blood and often attach themselves to people who enter the water.

Leek A thick-stemmed relative of the onion, but with a milder flavour. It is not bulbous like an onion but tall and cylindrical. It is a national emblem of Wales.

Legionary ant, *see* ARMY ANT

Legume A simple fruit that splits into two halves to release seeds (e.g. a pea-pod). The name is also given to any member of the true pea family, which typically possess legume fruits (pods).

Lemming Small rodents living in the cold regions of the northern hemisphere, where the vegetation consists mainly of lichens and stunted shrubs. They continue to live under the snow during the winter. Like the related voles, the lemmings periodically build up to huge numbers. They then emigrate in all directions.

Lemon A small evergreen tree related to the orange, producing large, tart, yellow fruits. The branches are spiny and the leaves are long and pointed. It is a native of the East Indies, but widely cultivated in warmer parts of the world, especially in southern Europe.

Lemur A group of primates confined to Malagasy (Madagascar). There are several species, ranging from 15 cm to one metre in length. In addition, they have long bushy tails. One of the best-known species is the ring-tailed lemur, which runs about with its black and white ringed tail held vertically.

Lenten rose, *see* CHRISTMAS ROSE

Lentil An annual herb of the Mediterranean region, related to the PEA. It has small spiny leaves and pale blue flowers. It bears edible seeds in short, broad pods. It is one of the oldest food plants.

Leopard One of the big cats, the leopard is a handsome animal up to 2.5 metres long, including the tail. The fur is basically tawny yellow, with black spots. Black individuals, known as panthers, are quite common. Leopards live in Africa and southern Asia, and feed on a wide variety of antelopes and other animals.

Lesser kudu, *see* KUDU

Lesser panda, *see* CAT-BEAR

Ring-tailed lemur

Three different types of lichen

encrusting

branched

leaf-like

Lettuce A hardy annual vegetable, probably of Asian or Mediterranean origin but only known as a cultivated plant. It has juicy green leaves growing close to the ground, which are cut and eaten before the plant shoots up to produce flowers. It is now the most popular salad vegetable and is a member of the Composite family.

Leucocyte The name given to the various white blood cells.

Lichen The seemingly lifeless yellow or grey patches that encrust rocks are extremely hardy plants called lichens. A lichen is really two plants in one — a combination of a fungus and an alga. The fungus absorbs water and produces acids that dissolve rock, releasing minerals. The algal cells use the water, the minerals and sunlight to make food for themselves and for the fungus. Such a mutually beneficial relationship is known as symbiosis. Lichens include branching, leaf-like and crust-like types. Large areas of the Arctic are covered with lichens known as reindeer moss.

Life cycle The sequence of changes which occur between the fertilized egg of one generation and the fertilized egg of the next generation, or from conception (the fertilization of the eggs) to death.

Ligament The tough strip of collagen fibres that joins bones together at a joint. Also the horny hinge of a bivalve mollusc shell.

Lignin The woody strengthening material of plants.

Lignum vitae A small tree of the West Indies and tropical America. It grows about 6 metres tall, and has leaves divided into two pairs of oval leaflets, and clusters of blue flowers. The wood is exceptionally hard, and sinks in water. It was called *lignum vitae* — wood of life — because its resin was thought to be a valuable medicine.

Ligule The 'collar' at the base of a grass leaf where it joins the leaf sheath. Also the strap-shaped projection of the florets of many composite flowers.

Lilac Cultivated shrubs of Asian and eastern European origin. The common lilac has heart-shaped leaves and spikes of purple or white flowers.

Lily The name of about 80 species of flowering herbs, and also of a family of around 8000 plants. True lilies grow from bulbs and have fleshy leaves and bell-shaped or trumpet-shaped flowers.

Lily of the valley A plant of the lily family, though not a true lily. It has white bell-shaped flowers on a long stem and oval leaves.

Lily of the valley

Lily trotter, *see* JACANA

Lime A tree of the orange family, producing green fruit pointed at each end. The tree originated in India, and is widely cultivated. Lime is also the name given to the linden, or basswood.

Limpet This name is given to various marine and freshwater snails in which the shell is more or less tent-shaped. When covered by water the animals glide about on the broad foot, and they scrape algae from the rocks. When the tide goes out the seashore limpets return to their own spots on the rocks and pull their shells down tightly around them. The rock and the shell are gradually worn so that they fit perfectly and it is then very hard to remove the limpet.

Linden, *see* LIME

Linear (of a leaf) Long and narrow, as a grass leaf.

Ling, *see* HEATHER

Linnet A shy finch with a beautiful song. It is found over much of the Old World, nesting in gorse and other dense bushes.

Linsang

Linsang A slender relative of the mongoose living in South-East Asia. The fur is short and velvety, brownish-grey with darker bands or spots. Linsangs are active at night, preying on a variety of small animals. They are excellent climbers.

Linseed, *see* FLAX

Lion Once found all over Africa and southern Asia, lions are now restricted to Africa south of the Sahara and a small area north of Bombay in India. Lions feed chiefly on zebra and antelopes.

Liquidamber The American gum tree, native to North and Central America, and three other species from eastern Asia. These large trees have lobed leaves, almost star-shaped, and yield fine resin. The timber is fine-grained.

Liquorice A herb of southern Europe, with compound leaves having many oval leaflets. It grows up to 90 cm tall, with blue pea-like flowers. Its roots are boiled to extract a food flavouring substance.

Litchi An evergreen Chinese tree. Its leaves are divided into lance-shaped leaflets, and it has clusters of small white flowers. The plum-sized red fruits have an edible, white, jelly-like pulp. The name is sometimes spelled lychee.

Littoral Inhabiting the shores and surrounding regions of the seabed.

Liver A large organ in the vertebrate body, closely linked to the digestive system and playing a major role in the management and distribution of food materials within the body. It acts like an amazingly complex warehouse and chemical laboratory. The following account applies mainly to the mammalian liver, but other vertebrate livers are much the same and differ only in some biochemical details.

Food digested in the intestine passes into the bloodstream and is carried straight to the liver, which stores or releases the materials as necessary. If the blood contains more GLUCOSE sugar than the body needs at the time, the liver removes some of it, converts it to a starch-like material called glycogen, and stores it. When the sugar content of the blood falls — during strenuous exercise, for example — some of the glycogen is re-converted to glucose and released. AMINO-ACIDS from the intestine are generally passed straight on to the tissues to be built up into new PROTEINS, but if there are too many the liver converts them into glycogen (which it stores) and urea. The latter is of no use and is returned to the blood to be removed by the kidneys and elimi-

Lions live in groups known as prides, and together they hunt the large herds of antelope and zebra which roam the African plains.

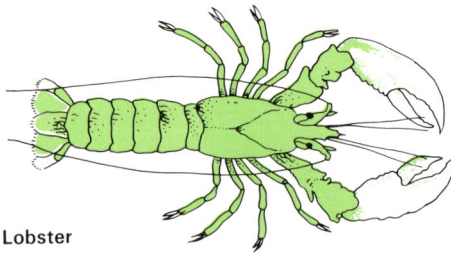

Lobster

nated from the body in the urine (see EXCRETION). The liver also stores fats and various vitamins and minerals, including iron. This is why animal liver is a useful food.

The liver's laboratory also destroys drugs and other poisons, such as alcohol, and makes them harmless. But if too much alcohol is drunk the liver cannot cope with it and becomes seriously damaged.

In addition to all these functions, the liver is a GLAND — the largest in the body — and its secretion, known as bile, is a mixture of various salts together with dark pigments from dead blood cells. The bile collects in the gall bladder, tucked up between the lobes of the liver, and flows down the bile duct and into the intestine after a meal. The bile is alkaline and helps to neutralize the acids from the stomach so that more digestive enzymes can get to work. The bile also acts rather like a detergent on the fatty food, breaking it down into droplets so that the enzymes can work on it more easily.

Liver fluke, *see* FLUKE

Liverwort One of 9000 or so species of plant which, along with the closely related mosses, make up the phylum Bryophyta. Liverworts may be green, brownish-purple or red and the stems and leaves are generally indistinguishable from each other. They grow mainly in damp and shady places, forming mats or cushions on the ground.

Living stone Any of a group of small succulent plants which live in deserts. They resemble small pebbles, but the body of the plant is actually a pair of thick leaves. The stem is underground. They produce yellow or white daisy-

like flowers, although they do not belong to the daisy family.

Livingstone daisy, *see* MESEMBRY-ANTHEMUM

Lizard Together with snakes, lizards make up the reptile order Squamata. Most lizards differ from snakes in having well defined limbs, though there are limbless forms, such as the SLOW WORM. Lizards also differ from snakes in having movable eyelids and external ears. Lizards are widely distributed throughout the world, particularly in the warmer regions. They range in size from a few centimetres to the 3-metre-long Komodo dragon. Most lizards lay eggs, but a few species give birth to live young. Lizards are generally carnivorous, although some eat plants.

Llama This South American relative of the camel is not really a wild animal, but a domesticated descendant of the guanaco. The llama has long been used as a beast of burden in the Andes.

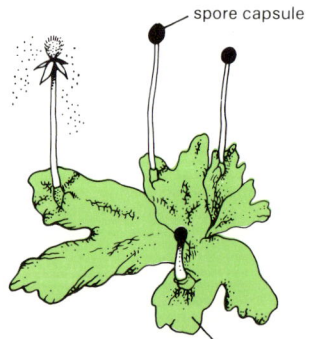

Liverwort   thallus

Loach Small freshwater fishes with cylindrical bodies found in Europe and Asia. Many have conspicuous markings, the best known being the orange and black clown or tiger loach.

Lobelia Any of about 250 species of flowering herbs or small shrubs. They occur in nearly all temperate or warm regions. They have tube-like flowers with flaring lips, and produce an acrid milky juice.

Lobster A large marine crustacean weighing as much as 18 kg. Lobsters live on the seabed and feed on a variety of living and dead animals. They have large claws and are blue when alive (turning red when boiled).

Locust Large grasshoppers whose populations periodically build up to vast numbers. At such times the insects fly out in great swarms and do a great deal of damage to crops. Africa suffers particularly from locusts, but various species are found throughout the warmer and drier parts of the world.

Locust tree, see CAROB

Loganberry A trailing plant similar to a bramble, believed to be a cross between a blackberry and a raspberry. It produces tart, dark red fruit like large raspberries. It originated in California in 1881.

Loggerhead A sea turtle widely distributed through tropical and temperate waters. It resembles a GREEN TURTLE with a large head.

London pride A small evergreen herb. It is a kind of SAXIFRAGE, originating in Spain and Portugal. It has a rosette of leathery leaves from which rise tall stems carrying star-like white and red flowers.

Longhorned beetle A group of beetles with long antennae — in one case four times the length of the body. The larvae of many species burrow in wood and are serious timber pests.

Loon, see DIVER

Loosestrife The name of two unrelated groups of plants, both growing in damp and marshy regions of Europe and Asia, and both bearing sprays of flowers on tall stems. Purple loosestrife is a common member of one group. Yellow loosestrife is a common member of the other group and is related to primroses.

Loquat A small evergreen tree from China and Japan. It has large leaves, downy underneath, and flowers in white drooping sprays. The yellow egg-shaped fruit has a slightly acid flavour. Loquat is a member of the rose family.

Lorikeet Colourful little parrots found from Malaya to Australia and the South Pacific. They are mainly green with other bright colours on the head and breast. Lorikeets live in flocks and feed mainly on nectar. They are also called honey-parrots.

Loris The lorises are slow-moving relatives of the lemurs living in the forests of South-East Asia. They are nocturnal animals with large forward-looking eyes. They have broad grasping hands and feet which can cling tightly to branches for long periods. Lorises have rather woolly fur.

Lory Closely related to lorikeets, but slightly larger, these parrots are mostly red in colour, with splashes of green and yellow. They live on the islands of South-East Asia and the south-western Pacific.

Lotus The name given to several different kinds of flowers. The Egyptian lotus and the tamara or sacred lotus of India and China are both kinds of water-lily. The name is also given to a large group of herbs, such as the bird's-foot-trefoil, related to peas.

Louse A small wingless insect which lives as a parasite on warm-blooded animals. Sucking lice are parasites of mammals and feed on the blood of their hosts. Biting lice live mostly on birds and feed on their feathers and skin. Man may be host to three kinds of lice, commonly called head, body and crab lice. Most serious is the body louse, a common carrier of typhus fever.

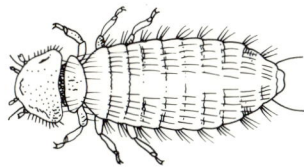

Biting louse

Lousewort A low-growing European herb growing in damp places. Louseworts have much-divided leaves and tubular, two-lipped flowers. There are many species, especially in uplands areas, all 'stealing' water and minerals from the roots of neighbouring grasses.

Lovage A member of the carrot family, lovage is a perennial herb of the north temperate zone. It has much-divided leaves, and the pink or white flowerheads are borne on tall stems. It is used for flavouring.

Lovebird A group of small colourful African parrakeets, generally green with patches of red or black on the face. Lovebirds pair for life and the pairs spend much of their time 'kissing' each other — hence their common name.

Love-lies-bleeding, *see* AMARANTH

Lucerne, *see* ALFALFA

Lugworm   castings

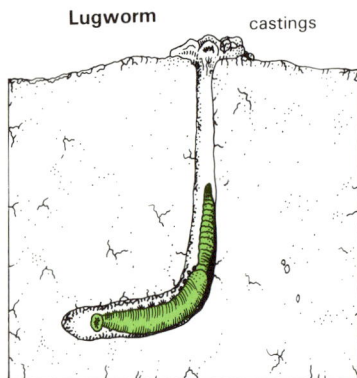

Lugworm A large worm living in U-shaped burrows in the sand of the seabed. Its coiled castings are a familiar sight at low tide. The animal is up to 30 cm long and looks rather like an earthworm. Lugworms swallow sand and extract food particles from it.

Lumpsucker A sluggish marine fish up to 60 cm long which spends most of its time clinging to rocks and seaweeds. Modified pelvic fins provide the fish with a powerful sucker.

Lung The air-breathing organ of mammals and all other vertebrates except the fishes. The name is also given to the breathing organs of terrestrial slugs and snails. In vertebrates there are normally two lungs (most snakes can accommodate only one in their slender bodies) and they are very thin-walled sacs in the front part of the body. They are little more than simple balloons in the frogs and toads and other amphibians, which breathe largely through the skin, but in the reptiles, birds, and mammals they are very complex. The lung surface is folded into thousands of tiny pouches, called alveoli, giving the lung a spongy texture and an enormous surface area. The surface area of a man's lungs is equivalent to about half a tennis court. Minute blood vessels surround each alveolus.

Each lung is joined to the windpipe or trachea by a tube called the bronchus, and the windpipe opens into the mouth and nose. When air is sucked into the lungs, oxygen passes through the thin walls and into the blood flowing around the alveoli. The oxygen is then carried round the body to be used in RESPIRATION. Waste carbon dioxide passes from the blood into the lungs and is expelled when the animal breathes out.

Lungfish A group of freshwater fishes found in South America, Africa and Australia. They have air-breathing lungs as well as gills, and they can live in very stagnant water. The Australian species has strong fins which allow it to walk about on the river-bed. The other species have eel-like bodies. The African and South American species can survive when the rivers dry up. They burrow into the mud and aestivate, surrounded by slime which keeps them moist.

Lupin Any of a group of American and southern European flowering plants. They have tall spikes bearing flowers somewhat like those of sweet peas, in red, blue, yellow or white. Tree lupins can grow up to 2.4 metres tall.

Lychee, *see* LITCHI

Lymph A colourless fluid that seeps out from the blood to feed and wash individual cells in the tissues.

Lymphatic system A network of large and small lymph-carrying vessels which branch throughout the body.

The small vessels penetrate every tissue except the nervous system and drain fluid (LYMPH) from between the cells, together with all particles too large to be absorbed by blood capillaries (e.g. colloids, bacteria, broken cell fragments). The lymph flows by muscular contraction of the vessels and flows into the vein system near the heart. Bacteria may be trapped and destroyed at certain swellings of the vessels — lymph nodes.

Lymphocyte A type of white blood cell made in the tissues of the lymphatic system, which helps to destroy bacteria, probably by producing antibodies.

Lynx Short-tailed, long-legged members of the cat family. They have tufted ears. The European lynx, up to one metre long, lives in the forests of northern Europe and Asia. It has a sandy coat and black spots. The Spanish lynx is a little smaller and more heavily spotted. The Canadian lynx is larger and has few spots.

Lyrebird The lyrebird is the largest of the perching birds. The male has a body about the size of a bantam and a distinctive lyre-shaped tail which gives the bird its name. Lyrebirds live in the mountain forests of Australia.

M

Macadamia A tropical evergreen tree of Australia. It grows more than 12 metres tall, and has leathery leaves and creamy flowers. It produces edible nuts (Australian nuts).

Macaque The macaques are the most common and widely spread of the Old World monkeys. They range from the mountains of North-West Africa to the forests of South-East Asia. Among the most familiar are the BARBARY APE and the rhesus monkey. The latter has been used extensively in biological research.

Macaw Large colourful parrots of tropical America. They have hooked beaks, and tails up to 60 cm long. Macaws live in large flocks, except at the breeding season. They feed on fruit and seeds, cracking hard nuts with their strong beaks and cleverly extracting the seeds with their tongues.

Mace, *see* NUTMEG

Mackerel A streamlined marine fish about 20 cm long which is widely caught for food. The common mackerel is found in the northern Atlantic and Pacific oceans.

Madder Any of several evergreen trailing plants of Europe and Asia related to BEDSTRAW. They have prickly stems and leathery leaves and small yellow flowers. A red dye was prepared from the roots in the past.

Lyrebird

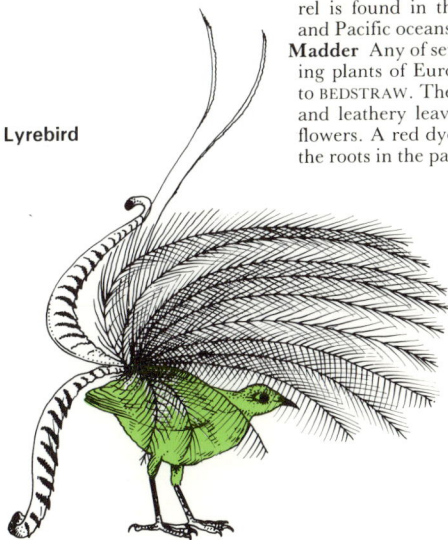

Magnolia Any of 75 species of ornamental trees and shrubs of Asia and North America. They range in height from 60 cm to 24 metres. They have sweet-smelling, showy flowers ranging from white to purple, or yellow.

Magpie A member of the crow family widely distributed over Europe, Asia and western North America. Measuring 45 cm in length, including a 25-cm tail, it is a conspicuous bird with iridescent black and white plumage. Magpies feed on a variety of small animals, particularly insects. They are also carrion-eaters and notorious egg-stealers.

Mahogany The name of a group of tropical American trees. They grow up to 30 metres tall. The leaves are divided into paired, glossy leaflets, and the flowers are yellow in clusters. The wood is hard and red. Similar looking timbers, especially from Africa, are often called mahogany.

Maidenhair tree. *see* GINKGO

Maize A type of cereal GRASS, producing fruit in cobs — grains packed around a woody core. It is called corn in North America. Maize is used as food for humans and animals, and for a wide range of industrial products. The grain contains large quantities of oil.

Maize moth Known in North America as the corn-borer, this moth is a major agricultural pest. Its larvae burrow into the stems and ears of maize plants and can ruin entire crops. It was accidentally imported into the United States from Europe early in this century and quickly spread through the vast cornfields of the Mid-West.

Malay bear, *see* SUN BEAR

Mallard A common wild duck of Europe, Asia and North America. For most of the year the drake is brightly coloured, with a glossy green head and neck separated from a rich brown breast by a white ring. In July the male moults his colourful feathers and for some weeks has a mottled brown plumage similar to that of the female.

Mallee fowl A turkey-like bird of Australia which incubates its eggs in pits covered with a pile of sand and decaying vegetation.

Mallow Any of a group of herbs of Europe, Asia and North Africa. The herbs have hairy stems and leaves, and showy red, pink or white flowers. The fruits are disc-shaped.

Malpighian layer The basal layer of the epidermis where the new skin cells are made.

Mandarin ducks

female   male

Mamba The African mambas are among the most dangerous snakes in the world. The black mamba is about 3 metres long, while the green mamba is usually less than 2 metres long. Black mambas live on the ground and are the fastest of snakes. Green mambas live in the trees and are less aggressive than the black species.

Mammal A warm-blooded vertebrate of the class Mammalia which nourishes its young with milk produced from the mammary glands. Mammals

Mallard

female   male

Manatee

are usually covered with hair but possess a self-regulating body temperature system which allows them to survive in areas of intense cold. The young are born alive with the exception of the egg-laying monotremes.

Mammary gland A milk-producing gland, characteristic of female mammals, used for suckling the young. The structure is probably a modified sweat-gland.

Manatee A small group of seal-like mammals, up to 4.5 metres long. One species lives around the West African coast, another species lives around the Caribbean coast. A third lives in the Amazon and Orinoco rivers in South America. All feed on aquatic vegetation.

Mandarin Any of several kinds of CIT-RUS trees producing fruits like small, flattened oranges. They include clementines and tangerines.

Mandarin duck A dabbling duck native to China which has been introduced into many parks. The male is brightly coloured with orange side-whiskers.

Mandible The lower jaw of vertebrates. Also the paired biting and crushing appendages near the mouth of insects and crustaceans.

Mandrake A perennial herb of the nightshade family, native to Mediterranean countries. It has large, lance-shaped leaves and blue or whitish bell-shaped flowers. The root is carrot-shaped and often forked.

Mandrill This is a forest-living baboon. The male's face and rear end are decorated with naked patches of pink and blue. The rest of the body is dark brown, with a golden collar and beard. Mandrills live in Central and West Africa. They spend most of their time on the ground, but sleep in the trees.

Maned wolf A long-legged fox of South America. It stands 75 cm at the shoulder and reaches 1.25 metres in length. The maned wolf is possibly the fastest member of the dog family.

Mangabey A group of tree-living, fruit-eating monkeys found in western and central Africa. Most are between 45 cm and 60 cm long, with a tail half as long again.

Mangel-wurzel A kind of BEET, used chiefly for cattle feed.

Mango A tropical evergreen tree, native to south-eastern Asia and now grown throughout the tropical regions. It grows up to 15 metres tall, with long slender leaves and clusters of tiny pink blossoms. It has oval stone-fruits, up to 30 cm long, with soft edible flesh.

Mangrove A tree that grows in tropical coastal swamps. It has thick, leathery leaves, and tangled stilt-like roots. It needs salt water to survive. Some produce edible fruits.

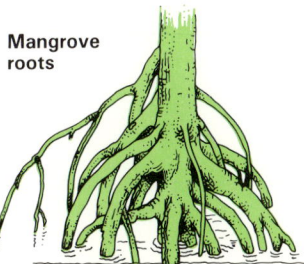

Mangrove roots

Manila hemp A fibre obtained from the abacá, a large plant of the BANANA family. It is obtained from the bases of the leaves which form the trunk.

Manna The sticky resin of the tamarisk. It is released when insects puncture the stem, and hardens to form white flakes. The name is also given to a similar substance exuded from the twigs of the flowering or manna ash tree.

Man-o'-war bird, *see* FRIGATEBIRD

Manta ray, *see* DEVIL FISH

Mantis A large group of voracious insects related to cockroaches. Mantises feed on other insects which they catch with their large spiny front legs. Mantises are often called praying mantises because, while waiting for prey to appear, they hold their front legs folded in a praying attitude. Most species live in the tropics but there are several in southern Europe.

Praying mantis

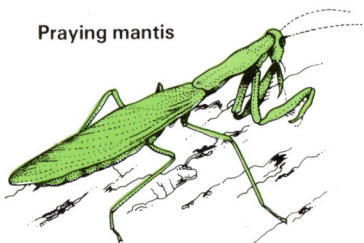

Mantis shrimp A marine shrimp with a pair of claws for seizing its prey. Ranging from 3–30 cm in length, mantis shrimps are found particularly in tropical waters.

Mantle A thick fold of skin covering all or part of the mollusc body. The mantle cavity, between mantle and body, contains the gills in most aquatic forms or else functions as a LUNG. The outer edge of the mantle secretes the shell.

Maple Any of 150 species of deciduous trees and shrubs of the north temperate zone. The leaves are broad and lobed. The fruits grow in winged pairs. The sap is rich in sugar, and the timber is hard. The sycamore is a maple.

Marabou A large stork of Central and East Africa. The back, wings and tail are dark grey, while the underparts are white. The head and neck are pink and more or less naked, with a fleshy pouch dangling from the throat. The bill is large and heavy.

Marestail A perennial herb growing in northern ponds and slow-running rivers. It has jointed stems with whorls of slender leaves, and very small green flowers.

Marigold The name of several unrelated plants. Most marigolds have orange-yellow flowers. Pot marigold, used for flavouring, comes from southern Europe; French and African marigolds are native to Mexico. All belong to the Compositae.

Marijuana, *see* HEMP

Marine iguana The only true marine lizard is found in the Galapagos Islands. It grows up to one metre in length and has a crest running the length of its body. Marine iguanas normally eat only marine algae.

Marjoram The name of a group of aromatic herbs, originally from the Mediterranean region. They have small downy leaves and purple flowers, and are used for flavouring food. Marjoram is a member of the mint family.

Markhor A wild goat of the Himalayan foothills. Males stand about one metre at the shoulder and have fine spiralled horns. The hair is sandy coloured in summer and grey in winter. Both sexes have a long beard and heavy mane.

Marlin Probably the fastest of all fishes, marlins reach speeds of 80 km/h. Their streamlined bodies — up to 4 metres long — end in a spear-like bill which can pierce the timbers of a boat.

Marmoset These squirrel-like South American animals are the smallest of the monkeys. There are several species, most about 20 cm long with bushy, 30-cm tails. They have claws instead of the more usual nails, and these help them to grip as they scurry through the tree-tops.

Marmot Mountain-dwelling rodents living in North America and Eurasia. The alpine marmot of Europe is about 50 cm long. Marmots live in small family groups, feeding on grasses and other plants. They hibernate in burrows during the winter.

Marram grass A grass with long creeping rhizomes which grows in sand-dunes in Europe and North Africa. It binds the sand and helps to stabilize the dunes. The stems are up to 130 cm tall.

Marrow A creeping plant, related to CUCUMBER; often called vegetable marrow. The leaves are large and hairy, and there are separate male and female flowers, both yellow and funnel-shaped. The fruits are large and cylindrical. Small varieties are called courgettes or zucchini.

Marsh A region of wet ground supporting typical vegetation such as marsh-marigold, rushes and water mint. The soil is of mineral matter and, unlike a fen or bog, there is no accumulation of peat.

Marsupial A pouched mammal. Marsupials bear their young alive but in a small, relatively unformed state. The female normally has a pouch on the lower side of the abdomen in which the young are carried and suckled until they are able to fend for themselves. Most marsupials live in Australasia but some are found in the Americas. Typical examples are the KANGAROOS, WALLABIES, OPOSSUMS and KOALA.

Marsupial anteater, *see* NUMBAT

Marmoset

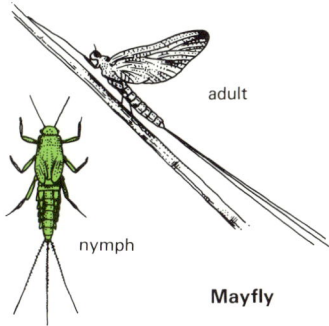

Mayfly

Marten A relative of the badger and stoat with a long body and short legs. There are several species in North America and Eurasia. The largest is the fisher marten, or pekan, about one metre long. This species hunts mainly on the ground, but the others live more in the trees and hunt squirrels and birds. Some of the species, notably the SABLE, have valuable fur.

Martial eagle The largest of the African eagles, with a wingspan of up to 2.5 metres. These birds live on the grasslands south of the Sahara and prey on a variety of small animals.

Martin A group of insect-eating birds related to swallows and migrating northwards with them each spring. Most easily distinguished from a swallow by its white rump, the martin catches its food on the wing. Most familiar is the house martin. Others include the American purple martin, and the sand martin which is found in most parts of the northern hemisphere.

Mastic tree An evergreen shrub of the Mediterranean region which exudes resin. The mastic resin is tapped by cutting the bark, and is used in varnishes and drugs.

Maté A South American variety of HOLLY. Its leaves are dried and used to make a drink, called Paraguay tea.

May, *see* HAWTHORN

May bug, *see* COCKCHAFER

Mayfly A group of insects whose nymphs live in water. The adults have one or two pairs of delicate wings and two or three long filaments on the end

145

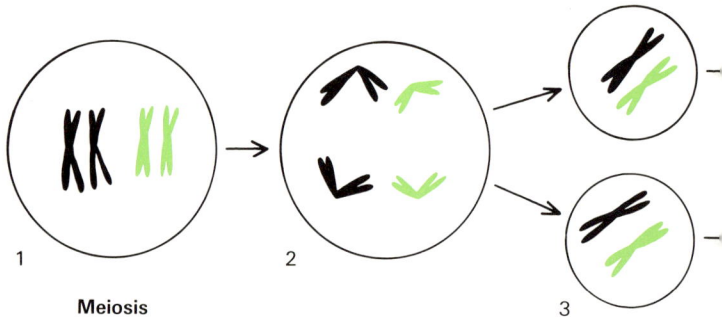

Meiosis

of the body. The adults do not feed and live for only a short time — a few hours in some species.

Meadow saffron, *see* COLCHICUM

Meagre, *see* CROAKER

Mealworm beetle A beetle which is better known in the larval form as a pest of stored grain and flour.

Medlar A small deciduous tree of the rose family, native to south-eastern Europe and Asia Minor. It has lance-shaped leaves with downy undersides, and large white flowers. The brownish fruit is not edible until it is about to decay.

Medusa The free-swimming saucer-shaped stage of the COELENTERATE life-history. The larger ones are the typical jellyfish which almost always reproduce by producing sex cells, although they are themselves formed by BUDDING.

Meiosis The process by which a cell nucleus divides when GAMETES or reproductive cells are being formed. The HOMOLOGOUS CHROMOSOMES pair up near the centre of the nucleus and one of each pair then moves to each end. The nucleus divides into two, followed by the whole cell, and each of the two daughter cells then has only half the original number of chromosomes. The cells usually divide again by MITOSIS to produce four gametes, each still with only half the original chromosome number. When sperm and egg cell unite at fertilization the full number of chromosomes is regained.

Melanism An excess of the pigment melanin in the tissues of organisms giving rise to black (melanic) individuals.

Melon A trailing plant of the GOURD family, producing large globular fruit full of juicy flesh and many seeds.

Menhaden A member of the herring family about 40 cm long. It is found in large numbers off the Atlantic coast of the United States and is an important food fish.

Menstrual cycle The cycle of changes in the female reproductive system found only in apes, human beings and the Old World monkeys which ends in the sloughing (shedding) of the uterine lining with some loss of blood. Menstruation takes place regularly during approximately 28-day cycles in human beings.

Merganser Colourful ducks with saw-edged bills. There are several species, breeding in freshwater and spending the rest of the year around the coasts. The commonest species, the red-breasted merganser, breeds throughout the northern latitudes and moves south for the winter.

Merlin A small falcon widely distributed on the hills and moors of the northern parts of North America and Eurasia. The male has a grey back and is smaller than the female. The female can be confused with a kestrel but is much darker brown and has a grey and black tail.

Mesembryanthemum A large group of South African plants related to the

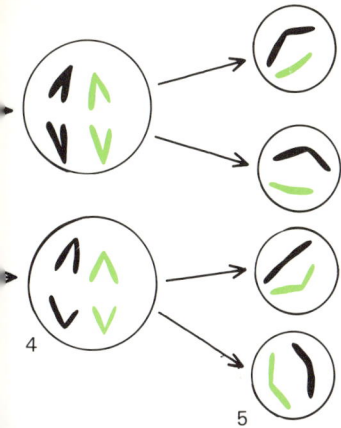

During the early stages of meiosis each chromosome makes a copy of itself, or duplicates, though the two halves remain attached to each other. Chromosomes then pair up around the centre of the cell nucleus (1). One of each pair moves towards opposite ends of the nucleus (2). The nucleus and the cell then divide into two daughter cells, each with only half the normal chromosome number (3). The chromosomes once again collect at the centre of the nucleus and the two halves finally split, with each half moving to opposite ends of the nucleus (4). The nuclei and then the cells divide into two again to give four gametes, each with only half the normal chromosome number (5).

LIVING STONES. They have succulent leaves and pink or white daisy-like flowers. They are often called LIVING-STONE DAISIES, although they not true daisies.

Mesentery Fine sheet of tissue suspending the stomach and other organs in the body cavity. Also the vertical partition found in some numbers in the gut cavity of corals and sea anemones.

Mesquite A low, thorny shrub of the American deserts. Very long roots enable them to reach permanent water deep in the ground and the shrubs remain green even in the driest seasons. Some specimens grow into trees up to 15 metres tall if there is abundant water deep in the ground.

Metabolism The sum total of processes — respiration, photosynthesis, etc. — that go on in an organism.

Metamorphosis Meaning a change of form, this term is applied to the marked and often very striking transformations which some animals undergo as they grow up. The change from a tadpole to a frog is a good example. Many crustaceans and most insects also show some degree of metamorphosis. It is not very marked in insects such as grasshoppers and earwigs, where the youngsters change gradually into the adult form and their wings develop gradually on the outside. This is called partial metamorphosis. Butterflies and moths,

however, undergo a much more startling transformation, from crawling, chewing caterpillars with no trace of wings to the flying, nectar-sipping adults. This change is called complete metamorphosis and it takes place during a special stage of the life history known as the pupa or chrysalis.

Metazoan A term used for all animals other than protozoans → i.e. the many-celled animals.

Micro-organism Any tiny organism, e.g. bacteria, that can be seen only with a microscope.

Midge The name given to three families of small mosquito-like flies: the non-biting midges, the biting midges and the gall midges. The biting midges feed on the blood of other animals, including man, and can be a nuisance on still summer evenings.

Midwife toad This small toad, about 6 cm long, gets its name because the

Midwife toad

147

male assists the female when she is laying the eggs and then carries them around attached to his back legs until they hatch. Midwife toads live in western Europe.

Mignonette A small perennial herb with lance-shaped leaves. It has tall spikes bearing sweet-smelling yellowish-white flowers.

Migration A fairly regular, instinctive movement of animals. There is normally a seasonal basis and the animals move one way in winter, the other in summer. Fishes going to the spawning grounds, swallows moving north in summer, reindeer moving south into the forests in winter, are all examples. Random mass movements with no later return are called emigrations and are normally brought about by overcrowding. The arctic lemmings are famed for their emigrations. Ladybird and locust swarms are also emigrations.

Mildew A term often used to refer to any fungal growth that covers the surface on which it grows with fine hyphae. More strictly, however, mildews are certain types of ascomycete fungi that are parasitic on the leaves of higher plants.

Milk teeth The first, deciduous teeth of mammals, later replaced by permanent teeth.

Miller's thumb, *see* BULLHEAD

Millet The name of a group of grasses. Some are grown to make hay or fodder, others for their grain, used for human and animal food.

Millipede The millipedes are arthropods in which nearly every segment of the body carries two pairs of legs. Most of them are long and slender, but some species, known as pill millipedes, are short and fat. They look more like woodlice and they can roll themselves into a tight ball. Millipedes live in the soil and leaf litter, feeding on decaying vegetation.

Millipede

Mimicry A phenomenon whereby an animal derives benefit from resembling another species. For instance, if one species of insect is protected from its predators by evil smell, sting or warning colours, other similar-looking species will also derive benefit. The resemblance will be continued and improved by natural selection over many generations. But mimicry helps some predators too. The most remarkable examples are the assassin bugs which feed upon other insects. They are remarkably like their prey, even to the smallest detail; some species resemble stick insects while others look like mosquitoes or even, in one case, the praying mantis. The assassin bugs can deceive even human collectors.

Mimosa Any of a group of 400 trees, shrubs and herbs in the pea family. They have tightly packed heads of small flowers — yellow, white, pink or purple — and feathery leaves.

Mink A semi-aquatic relative of the polecat. There are two species: the American mink and the European mink. The two are very similar, the males being up to 60 cm long including the tail and the females about half the size. Wild mink are brown, but many colour varieties have been produced in captivity, for the mink has a valuable fur and is reared on large farms.

Minke whale, *see* RORQUAL

Minnow A member of the carp family, the tiny European minnow is perhaps the most common of all freshwater fishes. There are many species including the hardhead minnow of California which may reach one metre in length.

Mint The name of any of a family of more than 3000 herbs, but especially of a group of about 25 species. These have square stems, creeping roots, strongly-scented leaves, and whorls of purple flowers. They are used to flavour food.

Mistletoe Any of a number of parasitic plants that grow on the branches of trees. Mistletoe is an evergreen, with small yellowish-green leaves, yellow flowers, and white berries. It forms a bush that hangs down from the host tree.

Mite A group of tiny arachnids which are found in almost every habitat.

Migration

Monarch
butterfly

Seal

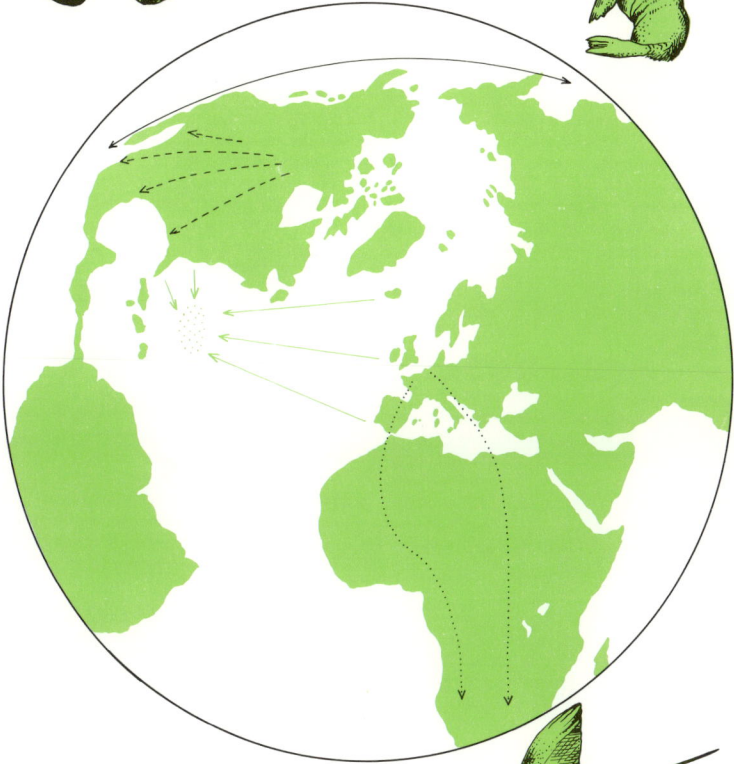

Eel

Swallow

In the summer monarch butterflies fly north to lay their eggs on milkweed
plants, the only plants on which their caterpillars feed. Every year thousands
of fur seals travel to a group of small islands off Alaska to breed. Swallows
fly south to the warmth of southern Africa during the cold European
winters. And strangest of all is the eel which travels across the Atlantic to
lay its eggs in the Sargasso sea.

Mitochondria

Most feed on decaying plant material but many are predatory. Others are external and internal parasites of other animals.

Mitochondria Tiny particles or organelles in a cell which carry all the enzymes involved in respiration and where the process takes place.

Mitosis The process whereby a cell nucleus divides so that each daughter nucleus — and therefore each new cell — receives *exactly* the same complement of CHROMOSOMES and GENES. This is essential if the new cells are to carry out their proper functions. At the beginning of mitosis the chromosomes gather near the centre of the nucleus and each then divides lengthwise into two identical halves. The two halves go to opposite ends of the nucleus, which then splits into two. The whole cell then splits into two, each new cell having an identical set of chromosomes and therefore identical features. Inside the young nuclei, the chromosomes fatten up again, ready for the next division; they gather in more DNA molecules and fit them together in exactly the right way so that each chromosome builds a replica of itself. The lengthwise division of the chromosome during mitosis is actually the separation of the replica from the original.

Mitten crab Small crabs which are natives of the rivers and estuaries of China and Japan. They get their name because the pincers are clothed with short hair and look as if they are wearing gloves. Mitten crabs reached Europe early in this century and now live in most of the rivers of western Europe.

Moccasin Properly called the cottonmouth water moccasin, this North American pit-viper may reach 2 metres in length. Cottonmouths are found in and around swamps and streams in the southern United States. This snake is quick to bite and possesses a potent venom which may result in death if not promptly and properly treated.

Mocking bird There are several mocking bird species in the Americas. They are about 30 cm long, with slender beaks and long tails. Mocking birds are popular creatures in the United States because of their ability to mimic other birds' songs. They can also make passable imitations of human voices and other man-made sounds.

Molar A crushing and grinding tooth in the cheek region of the mouth.

Mole Small mammals related to hedgehogs and shrews which are confined to the northern hemisphere. They spend almost all their lives under

During mitosis chromosomes duplicate but do not pair up as in meiosis (1). Each half of the chromosome then travels to opposite ends of the cell nucleus (2) and the nucleus and cell divides into two, with each daughter cell having the normal number of chromosomes (3).

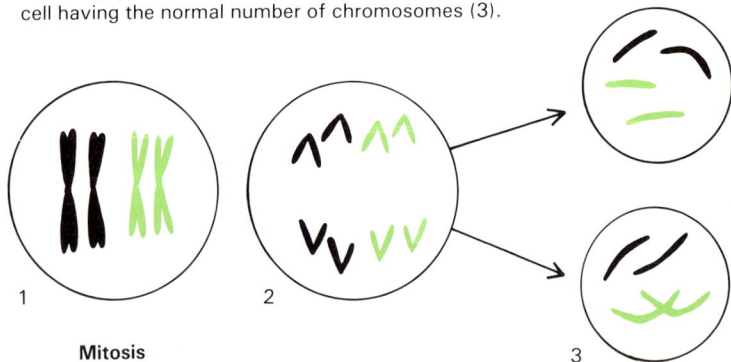

Mitosis

Mole

the ground and have very strong front limbs with which they excavate their tunnels. Their eyes are almost useless, and the animals rely on the senses of smell and hearing. They feed mainly on earthworms, but also take insects.

Moloch An Australian lizard which is covered with sharp spines. It is also called the thorny devil. The moloch is about 15 cm long and lives in the dry parts of the continent. It eats almost nothing but ants.

Mollusc A member of the phylum Mollusca which contains soft-bodied unsegmented animals, usually having a shell of lime salts. The shell may be external (e.g. bivalves) or internal (e.g. slugs). Molluscs have a definite head and a muscular foot on the lower side and usually breathe by means of gills. Examples include SLUGS, SNAILS, BIVALVES and OCTOPUSES.

Monarch A migratory butterfly. In North America monarchs make a remarkable migration between southern Canada and Florida or California — a round distance of 2000 miles — taking several generations to complete the journey. The monarch is also found in South America, Australia and New Zealand.

Mongoose Carnivorous mammals related to civets living in Africa and Asia. They have long tails, short legs and sharp muzzles. The largest is

about 1.2 metres long, almost half of which is the tail. Mongooses are very alert and quick animals, and they are well known for their ability to kill large poisonous snakes.

Monitor lizard The monitors include the largest of all living lizards. Several species exceed 1.5 metres in length and the Komodo dragon reaches a length of 3 metres. Monitors live in the warmer parts of the Old World, from Africa to Australia, and spend a lot of time basking in the sun. If disturbed these creatures can inflict painful wounds with their powerful jaws and lashing tails.

Monkey A term applied in its widest sense to any primate with the exception of lemurs and man. Monkeys may be subdivided into the narrow-nosed Old World monkeys and the broad-nosed New World monkeys.

Monkey bread, *see* BAOBAB

Monkfish A ray-like shark up to 2.5 metres in length found in many temperate and sub-tropical seas. It is a bottom-dweller, feeding on flatfish and other small creatures of the seabed.

Monkshood Any of about 100 species of perennial herbs, sometimes called aconites. It has deeply cut leaves, and its flowers look like monks' cowls. The drug aconitine is extracted from the roots of some species. All species are very poisonous.

Monocotyledon Any flowering plant that has only one seed-leaf. There are about 40,000, almost all with linear leaves with parallel veins. The flowers usually have 3 or 6 petals, although the grasses have none at all.

Monoecious Having separate male and female flowers, but both on the same plant.

Mongoose

Monotreme An egg-laying mammal. Examples include the PLATYPUS and the SPINY ANTEATER.

Moon rat, *see* GYMNURE

Moor Usually in upland regions, a COMMUNITY often dominated by certain tough grasses, heather and other acid-loving plants growing on damp or dry peat.

Moorhen Widely distributed in all temperate landmasses except Australia, the moorhen is never found far from water. It is a plump bird, about 30 cm long, with a brownish-black plumage and a distinctive red and yellow bill.

Moose The largest of the deer, standing 2 metres at the shoulder (although the rump is considerably lower). The male carries huge, flattened antlers. The moose lives in the northern forests of North America and also ranges from Scandinavia to Siberia. In Europe it is known as the elk. It is found mainly in the wetter regions, where there are plenty of willows for it to feed on. The animals also feed on water plants and often wade into lakes right up to their necks.

Moray eel These fishes live in the warmer seas of the world, especially around coral reefs. Some species reach lengths of 2 metres. The head and front part of the body are usually more bulky than the rest, and the mouth has numerous sharp teeth. There are no paired fins and no scales. Many are brightly coloured.

Morel Any of several species of edible fungi with wrinkled round or conical heads. Some similar species are not edible.

Morning glory Any of a group of climbing plants of the convolvulus family. Morning glories have heart-shaped leaves and funnel-shaped blooms in various colours from purple to white.

Morpho A group of large and brightly-coloured butterflies, usually blue, found in the tropical forests of Central and South America.

Morphology The study of external form.

Mosaic, *see* VIRUS

Mosquito Ranging from the tropics to the tundra, mosquitoes are found in many different habitats. Some species are blood-sucking pests of mammals, including man, and can transmit diseases such as malaria and yellow fever. Mosquitoes lay their eggs in water and most mosquito larvae 'hang' from the surface of the water, breathing through a respiratory tube which reaches the air above.

Moss A group of small spore-bearing plants that grow in dense clusters, forming bright green cushions on the ground. Moss also grows on walls and

Moose

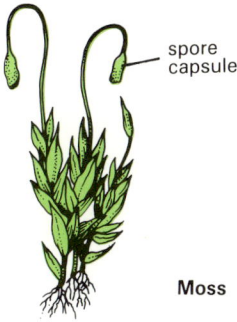

spore capsule

Moss

tree-trunks. The plants have short, slender leaves bearing thin, almost transparent leaves. There are no proper roots, although short, hair-like outgrowths anchor the plants to the ground.

Moss animal A group of colonial aquatic animals which form the phylum Bryozoa. Some are found in fresh water but most are marine. Many coat rocks and seaweeds; others resemble seaweeds themselves. Moss animals are often called sea mats. Each animal occupies its own cell in the colony and collects food by means of a ring of tentacles.

Moth A large group of insects which together with the butterflies make up the order Lepidoptera. Moths are mainly night-flying creatures and butterflies day-flying, but there are many exceptions to this rule. A better way of telling them apart is by the shape of their antennae. Butterflies have antennae with small knobs at the end. Moths have antennae of many shapes, but never with knobbed ends.

Motmot Colourful birds related to kingfishers living in Central and South America. Their beaks have saw-like edges, and most species have two long racquet-shaped tail feathers. Motmots are mainly insect-eating.

Motor nerve A nerve that carries impulses from the central nervous system to muscles and glands, stimulating them to action.

Mouflon This is the only wild sheep in Europe. It is about 70 cm high at the shoulder and reddish-brown in colour, with a whitish 'saddle'. The males usually have large curving horns, but the females are usually hornless. Truly wild mouflon are probably confined to Corsica and Sardinia, but flocks exist in the mountains of several countries.

Mould The name given to fungi whose hyphae or threads form a dense and often fluffy mat or mycelium over the food material on which it grows. Many damage food and cause disease in crops. *Penicillium* moulds produce the valuable antibiotic penicillin.

Mountain ash Any of several trees and shrubs of the northern hemisphere, mostly growing on high ground. Mountain ashes have compound leaves, white flowers, and scarlet or orange berries. Their other name is rowan.

Mountain hare Also known as the blue hare, the mountain hare has a white or blue-grey winter coat. It ranges from Scotland and Scandinavia across northern Asia to Japan. A closely related form, which rarely turns white in winter, is found in Ireland.

Mouse Any of a large group of smallish rodents with furry bodies and long, generally hairless, tails. The common HOUSE MOUSE is found throughout the world. The white mice used in laboratories or kept as pets are varieties of this species.

Mouse deer, *see* CHEVROTAIN

Mucor A very common fungus that grows on a variety of decaying organic

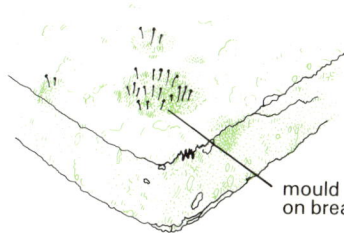

mould on bread

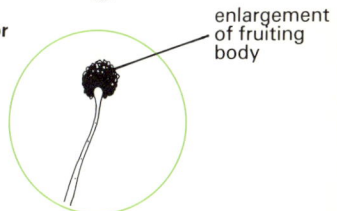

Mucor

enlargement of fruiting body

material — fruit, bread, etc. The tiny black fruiting bodies standing up above the white MYCELIUM have given it the name of 'pin-mould'.

Mudpuppy A North American salamander that never really grows up — it retains its gills throughout life and lives permanently in weedy ponds and streams.

Mudpuppy

Mudskipper A fish of tropical mudflats and mangrove swamps which is able to breathe air. It does not retreat with the tide but moves nimbly over the exposed mud, using its pectoral fins.

Mulberry Any of 10 trees with heart-shaped, toothed leaves, producing small blackberry-like fruit. The flowers are greenish-white. Silkworms eat leaves of the white mulberry.

Mule deer A common deer of western North America standing about one metre at the shoulder. It has a tawny coat, a white rump and a black-tipped tail.

Mullein Any of 100 species of biennial plants, related to foxgloves, producing a cluster of large leaves the first year, and a tall stem the second. Clusters of yellow flowers grow on this stem. The leaves of some species are very woolly or furry.

Mullet The name given to two groups of only distantly related marine fishes. Grey mullets grow up to 60 cm long and swim in shoals. Most species live in tropical waters, and some are important food fishes. The red mullet lives in tropical and temperate seas, particularly the Mediterranean, and is a valuable food fish.

Muntjac Small deer, 40–60 cm high at the shoulder and somewhat higher at the rear. Male muntjac have short antlers and long tusk-like canine teeth in the upper jaw. Muntjac come from South-East Asia, but several species have been introduced to other countries. They are now well established in England and live in wooded regions.

Muscle The tissue responsible for most animal movements, muscle is composed of slender cells called muscle fibres. Each is only about 0.1 mm across, but may be several centimetres long. The fibres are gathered into sheets or bundles. The most common kind of muscle in vertebrates is called striped muscle, because the fibres have a banded appearance under the microscope. It is sometimes known as skeletal muscle as well, because it is generally attached to the bones. It makes up almost all of the meat on an animal's body. Yet another name is voluntary muscle, because it is under conscious control, i.e. an animal can decide when to use this kind of muscle to make various movements.

The muscles work by contraction (shortening) of the fibres when they are stimulated by a signal from a nerve. This shortens the whole muscle which pulls on a bone or other organ, such as the eye-ball, and makes it move. The muscle cannot expand and push in the opposite direction, and so muscles have to work in pairs. The biceps muscle, which bends the arm at the elbow, works together with the triceps muscle at the back of the arm. When the biceps is contracted the arm bends and the triceps muscle is stretched. To straighten the arm again, the triceps contracts and pulls the arm straight, at the same time stretching the biceps. These skeletal muscles get tired quite easily and cannot stay contracted for long periods.

Mudskipper

Skeletal muscle

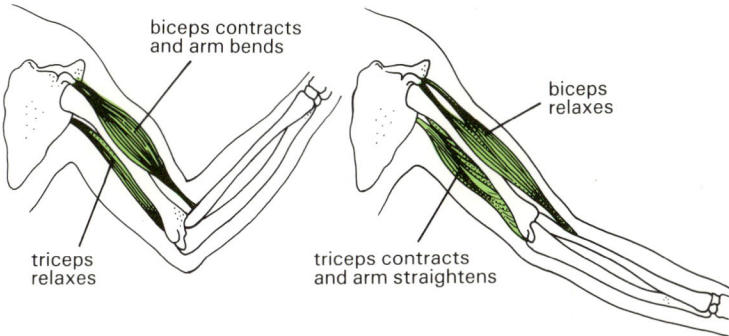

biceps contracts
and arm bends

biceps
relaxes

triceps
relaxes

triceps contracts
and arm straightens

Many of the muscles concerned with internal movements, such as those of the digestive system and the iris of the eye, are of a different kind, known as smooth muscle. The fibres are not banded. These muscles are not under conscious control and a person does not have to think about moving them; the autonomic part of the nervous system controls such movements. In addition, these muscles can stay contracted for long periods without getting tired.

Muscovy duck A large duck which has been introduced to many parts of the world from its original home in South America. Muscovy ducks tend to live on woodland lakes, for they are equally at home in trees as on water.

Musk ox This creature, more closely related to sheep and goats than to oxen, lives on the bleak tundra of Alaska, Canada and Greenland. The bull is a bulky animal, 1.5 metres high at the shoulder. The cow is nearly as large and both sexes have thick shaggy coats. Musk oxen have been introduced very successfully from Greenland to Norway and Spitzbergen.

Mushroom The name sometimes used for any umbrella-shaped fungus but more normally for a small group of edible fungi. Most familiar are the field mushroom and the closely related horse mushroom. The field mushroom has pink gills, turning purplish-brown with age. The horse mushroom has

white gills at first, turning purplish-brown with age and a stalk slightly swollen at the base.

Musk deer A small Himalayan deer until recently widely hunted for the musk secreted by the male. Musk is an important 'fixative' in perfume manufacture and is now made synthetically.

Musk rat A large vole up to 30 cm long, with a tail of about the same length. It is highly prized for its fur, known in the fur trade as musquash. Musk rats are natives of North America and when introduced to

Musk ox

155

Europe they did great damage to river banks.

Mussel The name applied to a number of bivalve molluscs. Best known are the shellfish that coat rocks along North Atlantic coasts. A large beard or byssus attaches each mussel to the rock. There are also numerous species of freshwater mussel.

Mustard The name of a number of annual herbs, and also of a whole family of plants related to cabbages. Mustard has large, rough jagged leaves, with stalks up to 90 cm tall, bearing yellow flowers. The hot seeds are used as a spice.

Mutant A gene or characteristic differing from the normal through having undergone mutation.

Mutation A sudden change in a gene or chromosome that leads to the appearance of new features in the organism possessing it. Because they are controlled by genes, these new features can be inherited. Most mutations are harmful and the animals do not normally survive, but some are useful and make the animal more successful. These useful mutations will be passed on to succeeding generations.

Mute swan, *see* SWAN

Mycelium The mass of tangled threads or hyphae of a fungus.

Mycology The study of fungi.

Mynah Large starlings of southern Asia. The Indian hill mynah, about 40 cm long, is a popular cage bird because it can mimic human voices. It is also known as a grackle. It is a glossy black bird with a yellow bill and yellow wattles behind the eyes.

Myosin, *see* ACTOMYOSIN

Myriapod A member of the arthropod class Myriapoda which contains the centipedes and millipedes. They have one pair of antennae, many pairs of legs and breathe by means of TRACHEAE. Centipedes have one pair of legs per segment while millipedes have two pairs per segment. Among many other differences, centipedes are carnivorous while millipedes feed on vegetable matter.

Myrtle An evergreen shrub of western Asia. It grows up to 3 metres tall, and has shining oval leaves and fragrant white flowers. The berries are purple.

N

Narcissus Any of a group of bulbous herbs with sword-shaped leaves. They produce flowers with six petals around a trumpet-like tube, white, yellow, or both. The DAFFODIL is a narcissus.

Nares Nostrils — the openings of the nasal or olfactory cavity.

Narwhal This is a smallish whale, reaching about 5 metres in length excluding the long tusk. The tusk is normally found only in the males and it is formed from one of the upper teeth. It may be 2.5 metres long. The narwhal lives in the Arctic Ocean and often forms great herds. It feeds mainly on cuttlefish and squids.

Nasturtium The name of about 50 species of trailing plants producing red, yellow or orange flowers. They are native to tropical America but widely cultivated in gardens. The rounded leaves are eaten as salad, and like the edible seeds have a spicy taste.

Native cat, *see* DASYURE

Natterjack The natterjack toad is easily recognized by the yellow line running down the centre of the head and back. Apart from that, it is very like the common toad. Natterjacks are found only in Europe. They are very local in the British Isles, but common in France and Spain. They live mainly in sandy places. They have very loud voices.

Nautilus The name given to two types of mollusc. One is the pearly-shelled nautilus of tropical seas which builds a spiral shell with numerous chambers, each bigger than the previous one. The animal lives in the largest and most recent chamber. The other is the paper nautilus or ARGONAUT.

Nectar A sweet, insect-attracting fluid produced by many flowering plants. Nectar is produced in nectaries which are placed in such a way that the insects have to brush against the stamens and stigmas in order to reach the nectar. The insects thus pollinate the flowers in the process.

Narwhal

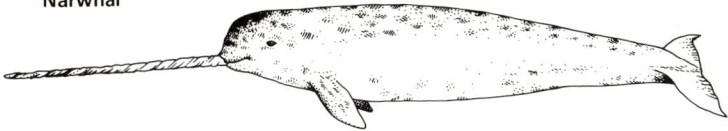

Nectarine A smooth-skinned variety of PEACH.

Nekton The free-swimming life in the sea as opposed to the floating plankton and the bottom-living benthos. Fishes, squids and whales make up most of the nekton.

Nematode A member of the phylum Nematoda which contains the round-worms or eelworms. These animals have cylindrical, unsegmented bodies, pointed at both ends. Nematodes are found in every conceivable habitat and as individuals are extremely numerous. Many are parasitic and cause severe damage to crops or to animals.

Ne-ne, *see* HAWAIIAN GOOSE

Nervous system This is the body's major control system, although the glands and hormones are also involved in controlling the body's activities. The very simplest animals have no nervous system, but in most animals it consists of a very complex network of nerves which carry signals or messages to and from all parts of the body. The brain and the spinal cord of vertebrates form the central nervous system, while all the other nerves form the peripheral nervous system, but the two are not really distinct: they are both part of one network and signals flow freely from one to the other.

The brain is the control centre, where messages are received from the eyes, ears, and all the other sense organs, including all the nerve endings in the skin and deep in the body. The brain is thus aware of everything that is going on inside and outside the body. It is like an amazing computer, sending instructions to all parts of the body and telling them how to respond to any situation. It can also store an immense amount of information for future use. The store of information is called memory.

Signals come into the brain from the sense organs, and are sent out to glands and muscles to produce action.

The signals travel along the nerves, which are bundles of very long cells or fibres, each insulated from its neighbour by a coating of fatty material. The signals are actually minute electric currents, and the insulation is necessary to prevent the signals from leaping across to the wrong nerve fibre. Each fibre is rather like a one-way street, carrying signals only in one direction, either from or to the brain.

Animals, including humans, have conscious control over many actions. They can decide whether to run or

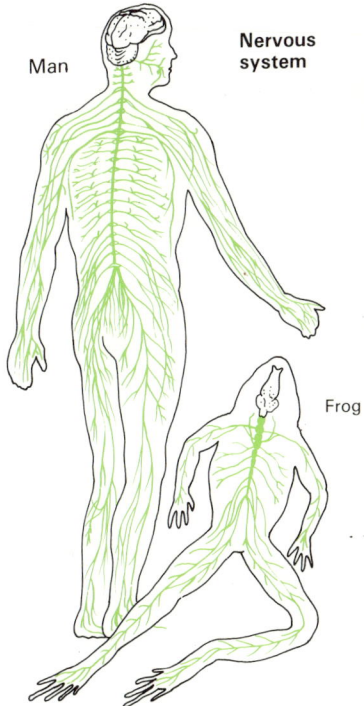

Nervous system

Man

Frog

walk, for example, and when to have something to eat. But there are lots of actions that are done automatically without having to be thought about. Such activities include breathing, keeping the heart beating and pushing food along the intestines. All these are controlled sub-consciously by part of the nervous system known as the autonomic nervous system.

Nettle Any of a group of plants with stinging hairs on their stems and toothed leaves. They produce small green flowers and have creeping roots. The stinging hairs discharge minute quantities of formic acid into the skin.

Newt The name given to a number of small lizard-like amphibians found in many parts of the world. Some adult forms live on land, some in water, but all lay their eggs in water. The smooth newt, the crested newt and the palmate newt are common European species. The spotted newt or red eft is a common North American species.

Nictitating membrane A third eyelid — a transparent fold of skin found in many birds and reptiles and some amphibians. Only a few mammals have a nictitating membrane.

Nidicolous birds Those hatching at an early stage of development and remaining in the nest for a relatively long period.

Nidifugous birds Those that hatch at a relatively advanced stage and leave the nest almost immediately — e.g. ducks.

Nightingale A small brownish bird with a beautiful song. Nightingales live in undergrowth and are rarely seen. They range from southern England to Asia in the summer but migrate to tropical Africa for the winter.

Nightjar The common nightjar, or goatsucker, is a nocturnal bird about 25 cm long. It is grey, brown and black and very difficult to see as it sits on the ground during the daytime. It produces a 'churring' noise throughout the night as it flies about catching insects on the wing. It is a migratory bird, spending the summer in Europe and Asia and going to Africa for the winter. There are many species of nightjar, all much alike, and they are found nearly all over the world. The whip-poor-wills are American nightjars.

Nightshade The name of a family of some 2000 plants, including petunia, potato, tobacco, and tomato. Some nightshades are poisonous, especially deadly nightshade, which has large oval leaves, purple flowers, and black berries.

Nile fish An eel-shaped electric fish of African rivers which swims forwards and backwards with equal ease. The Nile fish uses the electric field it generates to detect obstacles in the murky waters in which it lives.

Nile perch A valuable food fish of African lakes and rivers. It can reach one metre in length and may weigh over 120 kg.

Nilgai This is the largest Indian antelope, the males being nearly 1.5 metres high and weighing up to 270 kg. The male's horns are only about 20 cm long and the female is hornless. The male is bluish-grey and the female is tawny. These antelope can be recognized by the white 'garter' below each fetlock. Nilgai live in hilly country and open plains, feeding on trees and grasses.

Nitrogen cycle Nitrogen is one of the essential elements of life, being a major constituent of all proteins. In the form of nitrates, nitrogen is absorbed by plants and used to build up proteins. These are consumed by animals and

Night jar

converted to other proteins in the body. Upon the death of the organisms, the organic substances decay and bacteria convert the proteins back to nitrites and nitrates which can be used again by plants. This is the basis of the nitrogen cycle. A few bacteria, notably those forming nodules in the roots of leguminous plants, can convert atmospheric nitrogen into nitrates. Nitrates are also formed during thunderstorms: the energy of lightning causes oxygen and nitrogen to combine. The compound so formed dissolves in the rain water and falls to earth as a very dilute solution of nitric acid.

Numbat

Noctuid moth A large and varied group of night-flying moths, sometimes known as owlet moths. The smallest species has a wingspan of less than 1 cm; the largest spans over 30 cm. Most are dull coloured but some, especially those living in the tropics, have gaudy colours. Noctuid moths have 'ears' on the thorax which seem to be sensitive to the high-pitched squeaks emitted by bats — the main enemies of night-flying moths.

Noddy A small group of terns living in the warmer parts of the world. They are between 30 and 40 cm long. The white noddy, or fairy tern, is white, but the others are brown or grey. Most of them nest on remote islands, but they spend much of their time at sea. They catch small fish at the surface and do not dive.

Node A junction in a plant stem where leaves and branches may arise.

Norway haddock, see SCORPIONFISH

Notochord A flexible skeletal rod found at some stage during the life of all chordates. In most vertebrates it is found only in the embryo and is later replaced by the vertebral column.

Nucleic acid A complex compound which occurs in the chromosomes and in the cytoplasm of the cell. The genes are, in fact, chains of nucleic acid put together in various ways, and they control the features of the cells by controlling the types of protein that are made in the cells.

Nucleus The controlling body within the cell of an organism. The nucleus controls the activity of the cell by controlling the manufacture of proteins within the cell. The nucleus is also the seat of reproduction and it contains the CHROMOSOMES which contain the 'instructions' needed to produce new cells just like the parent cells. When a cell divides into two, the nucleus divides first and one new nucleus goes into each new cell.

Numbat A termite-eating marsupial, often called the banded anteater. It grows to a length of 45 cm, including a bushy 18-cm tail, and has brownish fur with white bands around the body. It has prominent ears and 52 teeth — more than any other land mammal. The numbat lives in south-western Australia.

Nut A true nut is a hard fruit containing a single seed. Examples include acorns and sweet chestnuts. Many of the 'nuts' we eat are not really nuts in the botanical sense. The walnut, for example, is just the woody inner layer of a stone fruit or DRUPE. The Brazil nut is a seed with a thick woody coat: there may be 20 or more of these seeds in a ball-shaped fruit.

Nuthatch A group of small birds with strong feet and claws which run up and down tree trunks with ease. They do

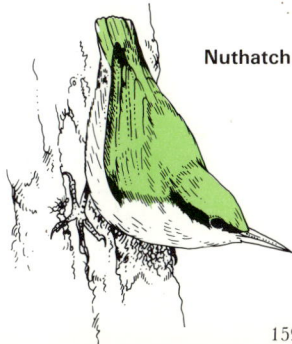

Nuthatch

not use their tails to support them. The birds eat insects for much of the year, but in autumn and winter they eat nuts, often wedging them into bark crevices and then hacking them open with their sharp beaks. Most of the nuthatches live in Eurasia.

Nutmeg An evergreen tree, originating in the East Indies. It has long pointed leaves, clusters of yellow flowers, and grows up to 21 metres tall. In the leathery fruit is a seed covered with a red membrane, the mace, used as a spice. The seed itself is also used as spice.

Nyala An ox-like antelope of southern Africa with white spots and stripes on its body. The males have twisted horns and weigh up to 125 kg.

Nymph The young stage of an insect that looks much like a miniature version of the adult.

O

Oak Any of 250 species of large trees, producing as fruit acorns, each partly enclosed in a cup. Some oaks are deciduous, others evergreen. Typical oak leaves are lobed and long. Oaks grow slowly and live for several centuries. They produce hard, strong timber.

Oarfish This marine fish, also called the ribbon fish, has a thin body, flattened from side to side, which may reach 6 metres in length. The deep red dorsal fin runs the whole length of the body and forms a spectacular crest over the head. The pelvic fins are long

acorn

Oak

and slender but broaden at their tips, rather like oars — hence the fish's name. Oarfishes live in nearly all parts of the oceans.

Oat One of the most important cereal GRASSES, widely grown for food. Oats have long branched stalks, with grain-producing spikelets not clustered into ears. They grow best in cool, moist climates.

Ocean sunfish Found in tropical waters, this fish has an oval body covered with a thick leathery skin. It may grow up to 3 metres in length and weigh over one tonne.

Ocelot This is one of the most beautiful members of the cat family. Its colour ranges from grey, through yellow, to deep brown, and it is decorated with spots and streaks. It is about one metre long and has rather long legs. Ocelots are forest-dwellers living in the warmer parts of America, from the southern United States to Paraguay.

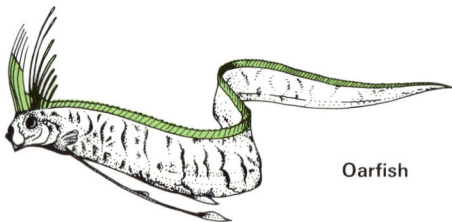

Oarfish

Ocellus The simple eye of many invertebrate animals, capable only of distinguishing light and dark.

Octopus A marine cephalopod with eight arms. These creatures range in size from a few centimetres across to as much as 10 metres. They feed mainly on crabs and shellfish which they catch with their tentacles, poison with venom and then open with their beak. Octopuses are found in all oceans of the world. They live on the seabed in nooks and crannies.

Oestrus cycle The cycle of changes in the female reproductive system of many mammal species. The cycle starts with the growth of the eggs in the ovary and the growth of the wall of the uterus. The eggs are then released and this is accompanied by a short period of oestrus or 'heat' during which the female is prepared to mate. If the eggs are fertilized, the cycle is suspended while the embryo develops. If fertilization does not occur, the lining of the uterus returns to its normal thickness. The cycle varies in time between species from a few days to several weeks. In many animals the cycle starts again immediately but some have definite breeding seasons and the oestrus cycle is active only at these times. Some animals have one oestrus cycle per year. The MENSTRUAL CYCLE is a variation of the oestrus cycle.

Oilbird A south American bird with the body of a nightjar, the face of an owl and the bill of a hawk. Oilbirds roost in caves during the day and emerge in the evening to feed on fruit.

Okapi

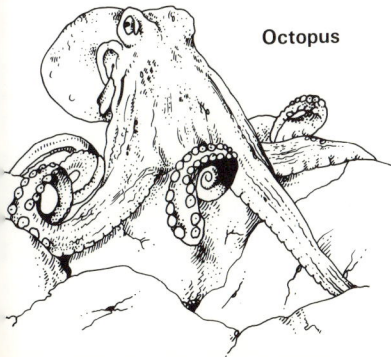

Octopus

Okapi This African mammal is the only close relative of the giraffe but the neck is not so long. The top of the head is less than 2 metres from the ground. Its coat is a rich reddish-brown and it is marked with white on the legs and haunches. Okapis live in the rain forests of central Africa and were not discovered by Europeans until less than 100 years ago.

Okra A tropical, hairy annual herb. It has heart-shaped, lobed leaves. It is cultivated for its long pods, eaten when young as a vegetable.

Oleander An evergreen shrub, also called rosebay, of the Mediterranean region. It has long, narrow, leathery leaves, and clusters of large red, pink or white funnel-shaped flowers. It is poisonous.

Olfactory sense This is the sense of smell. It is a chemical sense, with the sensitive cells actually stimulated by molecules of various substances floating in the air or dissolved in the water. Smell receptors are usually found on or in the head. Those of insects are usually concentrated on the ANTENNAE. Apart from the birds, most vertebrates have a good sense of smell. In fishes it is centred in pits on the snout. In air-breathing vertebrates the smell receptors occur in the nostrils and in the nasal cavity above the roof of the mouth. Each receptor is a single cell embedded in the lining, with a microscopic 'hair' reaching to the surface. It

161

is these 'hairs' that actually detect the scent molecules. They are continuously bathed by mucus flowing from glands in the nasal lining and the scent molecules must dissolve in this mucus before they can stimulate the 'hairs'. The mucus also washes away the old molecules so that the receptors are always ready to receive new ones. It is not known how the cells distinguish between different molecules and therefore recognize different scents.

Olive A small evergreen tree of the Mediterranean region. It has spiny branches, oblong leaves, and small white flowers. The fruit, like a small plum, yields a valuable oil.

Olm A blind amphibian that lives in underground rivers and lakes. It is related to the newts and salamanders but it does not grow up completely. It retains its red gills throughout its life. It has a white, eel-like body up to 30 cm long. Olms live only in the limestone caves of Yugoslavia and Italy.

Onion A small bulbous plant related to the LILY, and originating in Asia. The edible bulb is formed of leaves growing one inside the other. The plant is a biennial, and in the second year sends up a stalk with a rounded head of small white flowers.

Oogonium The female structure of some fungi and algae which contains the sex cells.

Oosphere A relatively large, non-mobile female gamete of many lower plants.

Oospore A fertilized oosphere with a thick outer wall — a resting stage.

Ooze A term given to many deep-sea deposits: e.g. radiolarian ooze, made up of the skeletons of tiny animals called radiolarians.

Opah An oceanic fish with an oval plate-like body up to 2 metres long. The fins are bright orange and the body has green and blue markings.

Operculum (1) The cover protecting the gill-slits of bony fishes. (2) The horny plate with which many gasteropods close their shell when at rest. (3) A lid covering moss capsules and fruits such as that of the poppy.

Opossum A group of American marsupials. The best known species is the Virginia opossum, common in many wooded parts of the United States and South America. This animal is up to 50 cm long and looks rather like a large rat. The young are carried in the mother's pouch for about 10 weeks and for a further month they ride about on her back. Not all opossums actually have a pouch.

Opossum shrimp A group of small shrimp-like crustaceans, the females of which have a brood pouch to hold their eggs. They are widely spread through the sea and fresh water.

Optic Concerning the eye.

Oral Concerning the mouth.

Orang-utan This is one of the apes. It lives in Borneo and Sumatra and is now in real danger of becoming extinct. The male stands about 1.3 metres when upright. The animals have rather sparse reddish-brown hair. They live in the forests and stay mainly in the trees, swinging from branch to branch with their arms. Orangs feed mainly on fruit.

Orang-utan

Orange A small citrus tree producing large, reddish-yellow fruit. It originated in Asia. It has dark evergreen leaves and white, waxy flowers.

Orb spider The collective name for those spiders which spin an orb-shaped web in which to catch their prey. The garden spider of Europe is a typical example. It is also known as the diadem spider from the jewel-like markings on its abdomen. When an insect is trapped in the sticky threads of the web the spider paralyses its prey with poisonous fangs and wraps it in silk.

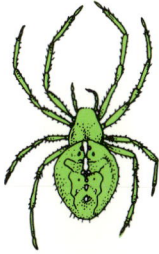

Orb spider

Orchid Any of about 17,000 species of perennial flowering plants, whose blooms consist of three petals and three petal-like sepals. The flowers are very irregular in shape, with one petal, called the tip, often much larger than the others. Some grow in the ground, others on trees as EPIPHYTES. Some orchids are SAPROPHYTES.

Order A grouping used in classification containing one or more related families.

Organelle Part of a cell forming a distinct unit which is specialized to perform a specific function within the cell, e.g. a FLAGELLUM, MITOCHONDRIA.

Oribi A small straight-horned dwarf antelope standing 60 cm high at the shoulder. It lives in open grassland over much of Africa south of the Sahara.

Oriole Orioles are forest birds living in the Old World. Only the golden oriole is found in Europe. This is a very striking black and yellow bird a little larger than a song thrush. The Old World orioles belong to a quite different family from the 'American orioles'.

Ormer, *see* ABALONE

Oryx The three species of oryx are among the most beautiful of the antelopes, with white or fawn coats and long slender horns. They all live in the desert areas of Africa and Arabia, the Arabian oryx being very nearly extinct in the wild. The scimitar oryx has distinctly curved horns and lives in the Sahara. The gemsbok has a fawn coat with black patches on the head and black stripes on the back and flanks. It lives in south-western Africa.

Osculum A large opening through which water leaves the body of sponges.

Osier The name given to several varieties of WILLOWS with long, flexible stems, used for basket-making.

Osmosis When two solutions, one more concentrated than the other, are separated by a semi-permeable membrane, liquid will flow from the less concentrated solution in an attempt to equalize the concentrations. This flow of liquid is called osmosis and the pressure which must be exerted on the more concentrated solution to stop the flow is called osmotic pressure. This is how plants suck in water from the soil (because sap is stronger than soil solution).

Osprey The osprey, or fish hawk, is a fish-eating bird of prey. Its narrow wings have a span of about 1.5 metres. It fishes in the sea and in lakes, diving

An epiphytic orchid

from a height of up to 30 metres and scooping up the fish with its talons. The head and underparts are whitish, but the rest of the plumage is brown. The osprey is found in all continents except Antarctica.

Ostrich The ostrich is the largest living bird. The male may stand up to 2.5 metres high, although half of this is taken up by the neck. The head, neck and legs are almost naked. The males have black plumage with white plumes on the wings and tail. The females are brown. Ostriches cannot fly, but they can run very fast. They live on the dry plains of eastern and southern Africa.

Otter A semi-aquatic carnivore related to the weasels. Otters are found in many parts of Europe. There are several species, all much alike. They have slim bodies, about 75 cm long excluding the tail, and short legs with webbed toes. The head is broad and flattened, and the ears are almost hidden in the sleek brown fur. Otters are powerful swimmers and they catch most of their food in the water. These animals frequently move from one lake or river to another and will often cover some distance on land.

Otter shrew A shrew of central Africa which looks like a miniature otter and leads a somewhat similar life. It has a

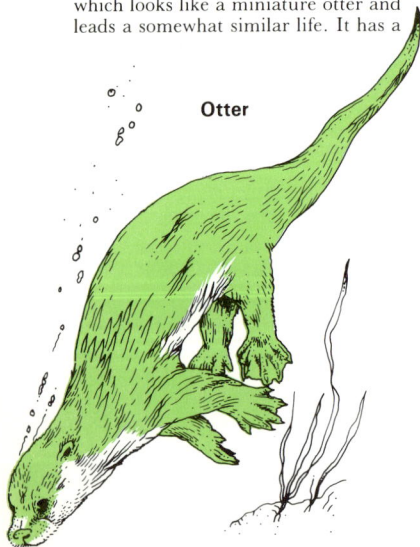

Otter

body up to 35 cm and a thick tail almost the same length.

Ounce, *see* SNOW LEOPARD

Ovary The organ in both seed-producing plants and animals in which female sex-cells are formed.

Ovenbird A group of South American birds named from the domed clay nests that some species make. Most ovenbirds live in Argentina and Chile. The North American ovenbird is not a true ovenbird and makes an oven-shaped nest, using grass not mud.

Oviduct The tube through which egg-cells pass from the ovary to the uterus. Fertilization may occur at some point in the oviduct.

Ovipositor An organ possessed by many female insects for placing eggs in a suitable position. The ovipositor is sometimes very long and in the ichneumon wasp *Rhyssa* it can drill a hole in wood to reach the grub in which *Rhyssa* lays its eggs. The stings of bees and wasps are modified ovipositors and that is why the drones cannot sting — they have no egg-laying apparatus.

Ovulation The release of a ripe egg-cell from the ovary.

Ovule A structure found in flowering plants and conifers that contains the female sex-cell and that develops into a seed after fertilization of the sex-cell. In flowering plants the ovule is protected within the carpel. In conifers and other gymnosperms the ovule lies naked on the scale of the cone.

Ovum (plural ova) Strictly an unfertilized egg-cell but the term is sometimes used for eggs — especially of insects.

Owl A nocturnal bird of prey found in most parts of the world. Their plumage consists of very soft feathers which make their flight noiseless. Owls have large eyes set in the front of the head and a keen sense of hearing. Among others they include the BARN OWL, SNOWY OWL and TAWNY OWL.

Owlet moth, *see* NOCTUID MOTH

Owl monkey, *see* DOUROUCOULI

Oxpecker A bird often seen perching on the backs of large African mammals. The oxpecker feeds on the ticks and flies of its host mammal while the host benefits from the removal of these parasites.

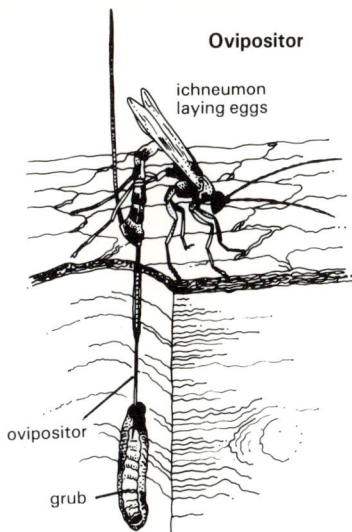

Ovipositor

ichneumon
laying eggs

ovipositor

grub

Oyster A group of edible bivalve molluscs. Once the food of poor people, oyster-beds have been so reduced through pollution and overfishing that oysters are now a luxury item.

Oystercatcher A large wading bird found along coasts in many parts of the world. The most common species has black and white plumage, a long red bill and pink legs. Oystercatchers feed in the tidal zone. Their diet includes cockles, mussels, worms and other shore creatures — but not oysters, which live below the low tide level.

Oyster plant, *see* SALSIFY

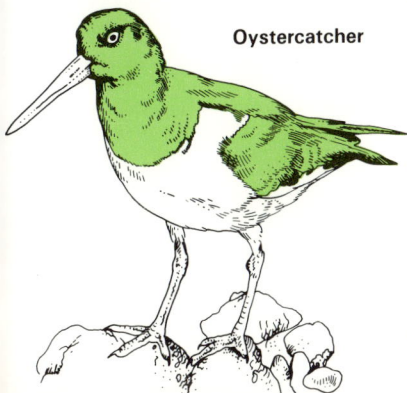

Oystercatcher

PQ

Paca A large rodent of Central and South America weighing up to 11 kg. Living in burrows near water during the day, pacas emerge at dusk to feed on fallen fruit and other vegetable matter.

Pack rat A group of voles living in North America. They are about 45 cm long, half of which consists of the tail. Pack rats are nocturnal animals found in almost every kind of habitat from woodland swamps to deserts.

Paddlefish A large freshwater fish related to the sturgeon. It has a long shark-like body with few scales and a long flattened snout looking rather like a canoe paddle. The snout may account for half of the fish's length. There are two species, one living in the Mississippi region of America, and the other in the Yangtze basin of China. The American species is almost 2 metres long, but the Chinese species is reported to reach a length of 7 metres.

Pademelon A group of small wallabies which live in thick scrub and forest undergrowth in New Guinea, Tasmania and coastal areas of eastern Australia. The largest measures only one metre in length, more than half of which is accounted for by its tail.

Painted finch, *see* GOULDIAN FINCH

Painted lady Known in North America as the thistle butterfly from the chief food plant of its larvae, the painted lady is a migratory butterfly widely distributed in both hemispheres. It is orange in colour with black markings and white spots on the forewings.

Painted turtle One of the most common and colourful turtles of North America. It has red stripes on the legs and yellow stripes on the neck and head. Painted turtles spend most of the time in ponds or the shallow parts of lakes, feeding on aquatic animals and vegetation.

Palate The roof of the mouth in vertebrates.

Palm Any of about 2500 species of tropical and subtropical trees. Palms have tall unbranched trunks ending in a crown of large fan-shaped or feathery leaves. The flowers grow in huge clusters. The fruits are berries or stone-fruits (drupes), either plum-like or with fibrous coats.

Palmate A palmate leaf is one with several lobes or leaflets radiating from a central point, e.g. horse-chestnut.

Palp A sensitive appendage on the head of various types of animal. Palps often play a part in feeding and are found especially in arthropods.

Pancreas An important vertebrate gland, or strictly a combination of two glands, which manufactures digestive enzymes and releases them into the intestine.

Panda The giant panda is probably the most famous of all animals. It is a black and white, bear-like creature, almost 2 metres long and weighing up to 140 kg. It lives in the cold, damp bamboo forests of Tibet and south-western China and is now carefully protected. Pandas eat mainly bamboo shoots, but they also eat other plants and various small animals incuding fishes.

Pangolin

Pangolin A mammal whose body is largely covered with overlapping horny plates. The seven species — four in Africa and three in Asia — are placed in an order by themselves (Pholidota). They have slender bodies and short legs, with long tails and snouts. Pangolins live in the forests and feed on ants and termites. They are often called scaly anteaters.

Panicle A type of inflorescence made up of a number of branched racemes.

Pansy A perennial garden plant of the violet family. It has a heart-shaped flower with a face-like pattern. The wild form is sometimes called hearts-ease.

Panther, *see* LEOPARD

Papaya, *see* PAWPAW

Paper nautilus, *see* NAUTILUS

Paprika, *see* CAPSICUM

Paradoxical frog The adult paradoxical frog, which lives in South America, looks much like any other frog. It is greenish-brown, with darker markings on the hindlegs. Its outstanding feature is the great size of the tadpole. When fully grown, the tadpole is more than 25 cm long, but then changes into a tiny froglet less than 5 cm long. The adult frog is only 8 cm long.

Paraguay tea, *see* MATÉ

Paramecium A single-celled freshwater animal, the largest being about 0.3 mm long. It is sometimes called a slipper animalcule because of its slipper-like shape. A paramecium swims by waving the hair-like cilia which cover its body. The animal normally reproduces by splitting in two (BINARY FISSION), a process that takes place several times a day.

Para nut, *see* BRAZIL NUT

Parasite An organism that lives in close association with another — often inside it — and takes food from it without giving anything in return. The organism that is attacked is called the host. Endoparasites are those that live inside their hosts. Examples include tapeworms, the organisms causing malaria and sleeping sickness, and the larvae of warble flies that burrow under the skin of cattle. Ectoparasites, such as fleas and lice, remain outside the host, although they may puncture the skin to suck blood. Many organisms are parasitic for only part of their lives.

Although originally believed to be a bear, the giant panda is now thought to be more closely related to the racoons. Despite its great bulk it is adept at climbing trees.

Parathyroid gland A gland which, in mammals, consists of four small spongy masses at the sides of the thyroid gland in the neck. These glands produce a HORMONE, called parathormone, which regulates the amount or calcium in the blood. The hormone helps to absorb calcium from food, and as long as there is plenty of this mineral in the diet the parathyroid glands remain small and not too active. If the diet lacks calcium, however, the glands swell up and produce lots of parathormone and cause calcium to be withdrawn from the bones. Parathyroid glands occur in other vertebrates as well, but are not always attached to the thyroid.

Parr, *see* SALMON

Partridges

Parrakeet Small brightly coloured parrots with long tails. The name has been given to a wide variety of species and does not refer to any clearly defined group. The budgerigar is the best known of the parrakeets.

Parrot The parrot family contains more than 300 different species, but only about one-third of them are strictly called parrots. The others are macaws, cockatoos, lorikeets, and so on. Parrots are generally stocky birds, with square or rounded tails. They all have large heads and strongly hooked beaks. One of the best known is the African grey parrot, which is a favourite pet because of its ability to mimic human voices. In the wild, parrots are forest-dwelling birds. They feed mainly on fruits and seeds.

Parrot fish Brilliantly coloured fishes in which the teeth are joined together to form a beak-like structure around the mouth. There are numerous species living around the coral reefs of the tropics. They feed mainly on seaweeds, but they also nibble away at the corals with their sharp beaks. Parrot fishes range from 30 cm to more than 3 metres in length.

Parsley A biennial herb of southern Europe with fine curled leaves. It is used for flavouring food. Parsley belongs to the same family (Umbelliferae) as the carrot and hogweed, and has small flowers carried in umbels.

Parsnip A vegetable related to the carrot. It has an edible, tapering white root. The leaves are very large compared with many other umbellifers.

Partridge A group of birds which include the grey partridge — the well-known game bird of Europe. The grey partridge has now been introduced to North America. Related birds include the red-legged partridge of south-western Europe and the tree partridges of south-eastern Asia.

Passerine bird A member of the order Passeriformes — the perching birds — whose first toe (hallux) is directed backwards and adapted for gripping branches. About half the living birds are passerines, including all the common garden birds — thrushes, tits, robin, etc. The crows are the largest passerines.

Passion flower Any of a group of American climbing plants with starlike flowers having five petals and five sepals. These flowers are often large and showy. Many species produce sweet edible fruit.

Patas monkey A long-legged ground-living monkey of the African savannas. Patas monkeys live in small troops which roam over a territory of about 50 sq. km, feeding on grass and berries.

Patella The knee-cap, a small bone in front of the knee-joint.

Pathogen A disease-causing organism, e.g. many bacteria and protozoans.

Pawpaw A small fast-growing North American tree with very large leaves and dark red flowers. It has sweet fruit looking like a small green or yellow rugby ball, although it is sometimes shaped more like a MARROW. Also called the papaya, the pawpaw is cultivated throughout the tropics.

Pea An annual climbing herb with hollow stems. It has compound leaves and tendrils which enable it to climb. The butterfly-shaped flowers are white or purple. The seeds are borne in pods, and are grown as food. The pea family includes alfalfa, beans and clover.

Peach A fruit tree closely related to PLUM. It is small to medium sized, with lance-shaped glossy leaves and small pink flowers. The fruit is roundish, yellow, containing a stone, and has a downy surface. Smooth peaches are called nectarines.

Peacock The Indian peacock, or more strictly peafowl, is the most familiar of three related birds living in central Africa and southern Asia. It has long been kept for ornamental purposes. The male has a magnificent train up to 1.5 metres long, which can be erected to form a shimmering greenish fan dotted with multicoloured 'eye-spots'. The peacock displays his train to peahens and to other peacocks alike.

Peacock butterfly A colourful butterfly found across Eurasia from Britain to Japan. It is easily recognized by the 'peacock eye' marking borne on each dark red velvety wing. The peacock

butterfly hibernates during the winter in dark sheltered places, such as barns and hollow tree trunks. Its larvae feed on stinging nettles.

Peacock worm, *see* FANWORM

Peanut An annual herb closely related to the pea, producing seeds in pods which ripen underground. It is also called a groundnut. Some peanut plants grow upright, others sprawl over the ground, rooting at the nodes. They produce yellow flowers. A native of South America, it is now grown all over the warmer parts of the world to provide valuable oil from the seeds, which are also eaten.

Pear A fruit tree closely related to the APPLE. It may grow up to 14 metres tall. It has oval, pointed leaves and clusters of white flowers. The fruit is fleshy and conical in shape.

Pearly-shelled nautilus, *see* NAUTILUS

Peat Partly decayed plant material that accumulates wherever water-logging or acidity slows down the processes of bacterial decay.

Pebble plant, *see* LIVING STONE

Pecan A North American tree, a kind of HICKORY. It produces large, woody stone-fruits with edible kernels.

Peccary The South American equivalent of the Old World pig. The tail is vestigial and the body is covered with thick bristly hairs. There are two species. The collared peccary ranges from Argentina to the south-western United States. It is greyish and has a paler collar of hair around the shoulder. The

Peafowl

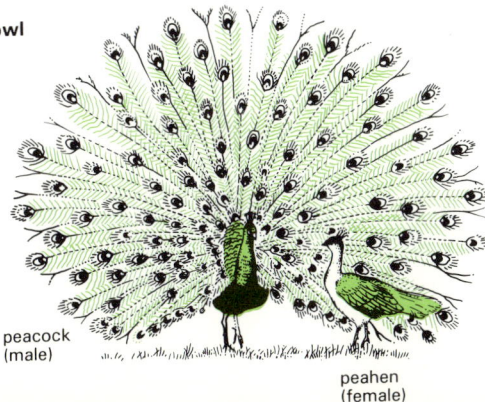

peacock (male)

peahen (female)

169

white-lipped peccary lives in the rain forests, never far from running water. It is reddish-brown or black, with a whitish area around the mouth.

Pedicel The stalk of an individual flower.

Peduncle The stalk of an inflorescence.

Peewit, *see* LAPWING

Pekan, *see* MARTEN

Pelagic Inhabiting the open sea but not fixed to the bottom. Pelagic animals are divided into free-floating PLANK-TON and actively swimming NEKTON.

Pelargonium The name of a group of small, shrubby perennial plants. They are native to South Africa and Australia. They have large leaves and showy flowers and are widely cultivated. They are related to the GERANIUM.

Pelican The pelicans are among the largest birds, some having wingspans of 3 metres. They are rather ungainly on the ground, but they are expert swimmers and fliers. The birds have long necks, and the large beak carries an extending pouch which is used for catching fish. The bird swims on the water and dips its head under the surface, using the pouch as a net. The brown pelican is a seabird, but the other species, which are mainly white, live around inland lakes. They are found in the warmer parts of the world.

Pelican

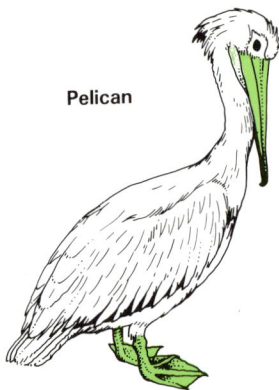

Emperor penguin

Peltate (of a leaf) Flat, with the stalk attached below — e.g. nasturtium leaf.

Penguin A flightless aquatic bird of the southern hemisphere, particularly of the Antarctic region. Penguins are truly at home in the water where they can swim at speeds of up to 40 km/h. On land they are extremely clumsy and slide on their bellies when there is a need to move rapidly. Penguins feed mainly on crustaceans and fish. The largest species is the EMPEROR PENGUIN.

Penicillium, *see* MOULD

Penis The male organ used during mating among higher animals to transfer sperms from the male to the female. In mammals it also carries the URETHRA from the bladder.

Peony Any of a group of perennial herbs native to Asia, southern Europe, and America. They have red and green shoots in spring, and very large blossoms, from white to crimson. Tree peonies form woody bushes.

Pepper The name of several plants producing pungent spices. Ordinary black and white pepper comes from a trailing shrub of the East Indies, with large leaves and tiny flowers. The fruits ripen to become peppercorns. Green, red and yellow peppers used as vegetables are varieties of CAPSICUM, as is CAYENNE PEPPER.

Peppermint A perennial herb, a variety of MINT, grown for its fragrant oil.

It has smooth oval leaves and pale blue flowers.

Perch The name of two closely related freshwater fishes, the common perch of northern Eurasia and the yellow perch of eastern North America. Both are gamefishes with dark striped sides and reddish lower fins. They are typical of a large family of spiny-rayed fishes.

Père David's deer This deer, which stands a little over one metre high at the shoulder, came originally from the plains of northern China, but it no longer exists in the wild. There is a herd at Woburn Park in England. The male carries large antlers which fork near the base into two more or less equal arms. The hooves are large and spreading, and it is believed that the animals originally lived in swampy areas.

Peregrine falcon

Peregrine A large falcon, up to 45 cm long. The upper surface is slaty blue and the underparts are white, both regions having dark markings. Peregrines are most numerous in rocky areas, especially around the coast. They are swift birds and feed mainly on other birds, especially pigeons. Peregrines live in all continents except Antarctica.

Perennation The survival of a plant from one year to the next by means of vegetative organs such as corms or rhizomes.

Perennial A plant that continues to produce flowers and seeds year after year.

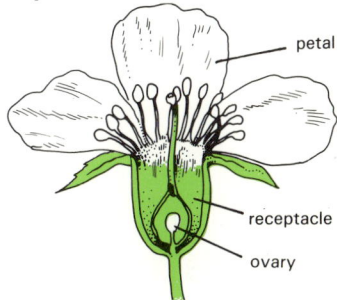

Perch

Perianth The sepals and petals of a flower, especially when, as in bluebells, these are all alike.

Perigynous flowers have a cup-shaped or dish-shaped receptacle with petals on the rim, around the outside of the stamens and carpels.

Peripatus A primitive worm-like arthropod which lives in the damp forests of the southern hemisphere. It was once thought to be an evolutionary link between the ringed worms and the arthropods, but zoologists now believe that it has existed with little change for at least 500 million years and it is possible that later arthropods evolved from worm-like ancestors through stages very much like the peripatus.

Peripheral nervous system, *see* NERVOUS SYSTEM

Peristalsis A rhythmic muscular contraction of the alimentary canal that causes food to pass along.

Periwinkle (botanical) A small trailing shrub, native to Europe. It has blue flowers and leathery evergreen leaves. There are several similar species, including one that flourishes in the tropics.

Perigynous flower

Periwinkle (zoological) The common name given to several species of sea-snail. The largest (the winkle of the fishmonger's shop) has a coiled shell over 3 cm in length and has long been collected from the seashore as a source of food.

Petrel A group of small ocean birds that normally come ashore only to breed. The storm petrel has a flitting flight and frequently follows ships far out at sea.

Persimmon Any of a group of small trees related to ebony. They have large, shiny leaves and red tomato-like fruit which is very sweet when fully ripe. Persimmons grow in most of the warmer regions.

Petiole A leaf-stalk.

pH An index figure denoting the degree of acidity or alkalinity of a solution: ph 7 is neutral; lower figures denote acidity and higher ones alkalinity.

Phagocyte A cell, especially of the blood, that engulfs particles from its surroundings in the manner of *Amoeba*. Blood phagocytes are very important in defending the body against bacterial infection.

Phalanger Tree-dwelling Australian marsupials which fill the niches occupied by monkeys and squirrels in other parts of the world. There are several species and they all feed mainly on fruit and leaves.

Phalanges Bones of the fingers and toes.

Phalarope A group of small wading birds with very slender bills and lobed toes. The summer plumage of the female is more brightly coloured than that of the male. Phalaropes breed in the northern parts of the world, migrating south in winter. The male incubates the eggs.

Pharynx The region at the back of the mouth into which the windpipe opens. In fishes the gill slits open into the pharynx.

Pheasant The name given to a number of birds but particularly applied to two species: the green pheasant of Japan and the game pheasant. The latter originated in Asia, but it has been naturalized in many countries. The female is dull brown but the male has a brilliant plumage and a long tapering tail.

Phloem The conducting tissue that carries sugars and other manufactured food materials around a plant.

Phlox Any of about 50 species of North American and Siberian herbs, grown for their salver-shaped flowers of red, violet or white. The blooms grow in clusters on the top of stems which may be up to 90 cm tall.

Photophore A light-producing organ.

Photosynthesis The process whereby green plants manufacture sugars from water and carbon dioxide using the energy of sunlight. No other organisms can do this and green plants are therefore the primary food producers in the world — all animals depend ultimately on plants, either by feeding directly on them or by feeding on other animals that themselves feed on plants. Photosynthesis takes place mainly in the leaves of the plant and only during the hours of daylight. Water is absorbed through the root of the plant and carbon dioxide is absorbed from the air through the stomata on the leaf surface. The green pigment chlorophyll plays a vital role in 'trapping' the energy of sunlight.

Phototrophic Feeding by photosynthesis.

Phylum The largest category used in the CLASSIFICATION of living organisms. All members of a phylum have a basic similarity of structure but may differ greatly in detail.

Pheasant (male)

Physiology The study of living processes, such as digestion.

Phytoplankton The minute plants that float in their millions near the surface of the seas and lakes.

Piddock A marine bivalve mollusc that bores into stone and wood. It has a rather delicate long shell. The hinge is so arranged that the two halves of the shell can rock backwards and forwards, and it is this motion that gradually bores a hole into the stone aided by rough teeth on the shell surface. The largest of the piddocks is about 15 cm long.

known as sardines, form the basis of an important canning industry in Spain and Portugal.

Pilot fish A small black and white striped fish that swims about in the company of whales, sharks and other large fishes. Pilot fish do not guide their larger companions in any way.

Pilot whale A large dolphin which lives in schools, often several hundred strong. It is found in all but the coldest seas. The habit of pilot whales blindly following a leader sometimes results in whole schools being stranded in small bays.

Pike

Pig Wild pig, or boar, range through western Europe, northern Africa and the southern half of Asia. They are the ancestors of domesticated pigs. A male (boar) can grow to almost 2 metres in length and weigh up to 200 kg. Its tusks may be 30 cm long. The female (sow) is smaller than the boar.

Pigeon An alternative name for dove. Some birds are normally known by one name, e.g. turtledove and some by both, e.g. rockdove or rock pigeon.

Pigweed, *see* GOOSEFOOT

Pika Small mammals related to rabbits and hares found in North America and Asia. They look like rabbits except that they have short rounded ears and all four legs are about the same size. There is no tail. Pikas live in a variety of grassy habitats from lowland plains to mountain slopes.

Pike A fierce freshwater fish, sometimes reaching 1.5 metres in length, with a flattened head and large jaws bristling with teeth. The fish is greenish, and it blends well with weeds among which it hides. When another fish swims close enough the pike will dart out and snap at it.

Pilchard A herring-like fish about 25 cm long which lives in huge shoals off the coasts of western Europe and in the Mediterranean. Young pilchards,

Pimento, *see* ALLSPICE

Pimpernel Small annual herbs of Eurasia and North Africa. Best known is the scarlet pimpernel, with a trailing square stem. It has oval leaves and small star-shaped flowers which close in dull weather. They can be red, blue or white.

Pine A group of about 90 species of large evergreen coniferous trees. They

Wild pig and young

have needle-shaped leaves growing in bundles, and bear separate male and female cones. The tallest is the sugar pine which can reach 75 metres. Pines are quick-growing and have soft timber.

Pineapple A South American perennial herb, producing a cluster of sword-shaped, sharp-edged leaves. It has a conical cluster of flowers which grow together to form the pineapple fruit, topped with small leaves.

Pine grosbeak, see GROSBEAK

Pine lizard, see FENCE LIZARD

Pine-marten A small but fierce carnivore of the northern forests of Europe and Asia, with a long body, a bushy tail and short legs. It is very fast and agile in the trees and often catches squirrels and birds, but actually hunts mainly on the ground. Voles and other small rodents are its main prey. It also eats some fruit.

Pine processionary, see PROCESSIONARY MOTH

Pink A tufted perennial European herb. The fragrant flowers have petals with crinkled edges in shades of pink, white or red. There are many cultivated species, including CARNATIONS and sweet williams.

Pin mould, see MUCOR

Pinna, see EAR

Pinnate A pinnate leaf is a compound leaf with a single main axis and a number of separate paired leaflets attached to it, e.g. ash and walnut.

Pintail A surface-feeding duck widespread over the northern hemisphere in summer. It migrates south for the winter. A pair of long tail feathers, especially marked in the drake, give this bird its name.

Pipefish A long thin fish with a long snout which often swims in an upright position through the water. The male pipefish keeps the eggs and the young in a pouch on his body. Slow swimmers, the pipefishes depend upon bony armour for protection.

Pipistrelle The commonest and smallest European bat. The wingspan is about 20 cm yet the total body weight does not exceed 7 grams. The pipistrelle ventures out earlier in the evening than other bats and can be recognized by its jerky flight.

Pipit A group of small insect-eating birds found in most parts of the world, usually in open country. They are generally brown in colour and some species are difficult to tell apart.

Piranha A ferocious fish found in South American rivers. It travels in schools and has been known to attack and devour large animals within a matter of minutes. Piranhas are between 5 and 20 cm long and have extremely sharp teeth.

Piranha

Pistachio A small deciduous tree, native to western Asia, with wide-spreading branches. It has resinous leaves and small brownish flowers, male and female on separate trees, and produces a small woody stone-fruit with an edible kernel.

Pistil, see GYNAECIUM

Pitcher plant Any of about 85 species of carnivorous plants which trap their insect prey in modified leaves which form pitchers. Rainwater collects in the pitchers. Insects attracted by honey glands fall into the pitchers and drown and are then digested by the plant.

Pitta A group of thrush-like birds living mostly in south-eastern Asia. The plumage is often brightly coloured in a rich variety of patterns.

Pituitary gland A small gland attached to the underside of the vertebrate brain and rightly called the master gland, although it weighs less than a gram in man! It is so named because it produces numerous HORMONES and affects almost every function in the body. It even controls the rate at which many of the other glands work. The pituitary gland is especially involved with the rate of growth and metabolism. If it is over-active in children it can cause them to grow into giants, and if it is not active enough it stunts growth and produces dwarfs. But these cases are very rare. The gland is sensitive to the amount of salt in the blood and sends out a hormone which 'instructs' the kidneys how much water to remove in order to maintain the right concentration. Other pituitary hormones affect the reproductive organs, and one of them controls the menstrual cycle in women (see MENSTRUATION).

Pit-viper A group of poisonous snakes found throughout North and South America and Asia. They have a characteristic deep pit on each side of the face and long movable front fangs which fold up out of the way when the mouth is closed. The pit is a sensitive heat receptor, capable of detecting the warm bodies of mammals and birds. Some Asian and South American species are small tree-climbing snakes while most are ground-dwelling. The largest species are the eastern diamondback rattlesnake of the south-eastern US and the tropical bushmaster. Both may be more than 2.5 metres long. The bite of a pit-viper is painful and, in some species, fatal.

Placenta A structure that develops in the UTERUS or womb of most female mammals when they become pregnant. It is formed largely from the lining of the womb and the developing baby (foetus) is attached to it by the umbilical cord. The placenta is full of blood vessels. Most of them are part of the mother's blood system, but some belong to the foetus. These two sets of blood vessels do not actually join up, but they mingle very closely in the placenta. Food and oxygen can therefore pass from the mother's blood into that of the unborn baby. The mother continues to feed her baby or babies in this way until they are born. The placenta comes away afterwards and is known as the afterbirth. Mammals that have a placenta are known as placental mammals. Pouched mammals or MARSUPIALS have no placenta. There is no way in which their babies

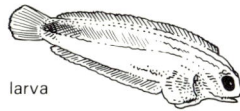

larva

larva

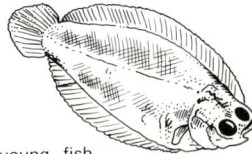

young fish

adult

Development of the plaice

can be fed inside the mother's body, and so they are born at a very early stage when they are very tiny. They continue their development in the mother's pouch, feeding on her milk.

Plaice Commercially the most important flatfish in the North Atlantic, the plaice habitually lies on its whitish left side. Its right side, which is always uppermost, is brown with red dots. Plaice are bottom-dwelling fish and spend much of the time slightly buried in sand or mud. Young plaice live in the surface waters, feeding on planktonic animals. They have the normal shape of fish larvae and swim upright. It is only as they develop that the body becomes flattened and the fish begins to swim on its side.

Plains hare, *see* JACK RABBIT

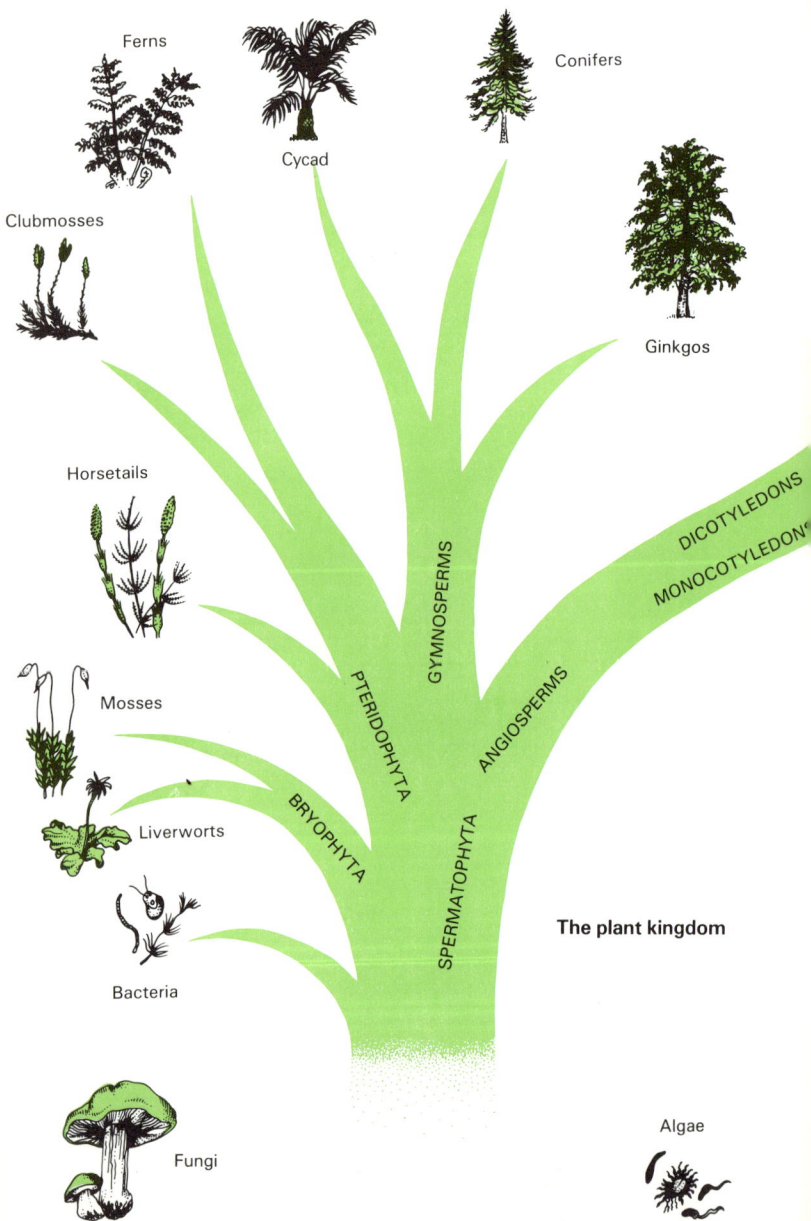

Ferns

Cycad

Conifers

Clubmosses

Ginkgos

Horsetails

DICOTYLEDONS

MONOCOTYLEDONS

GYMNOSPERMS

Mosses

PTERIDOPHYTA

ANGIOSPERMS

Liverworts

BRYOPHYTA

Bacteria

SPERMATOPHYTA

The plant kingdom

Fungi

Algae

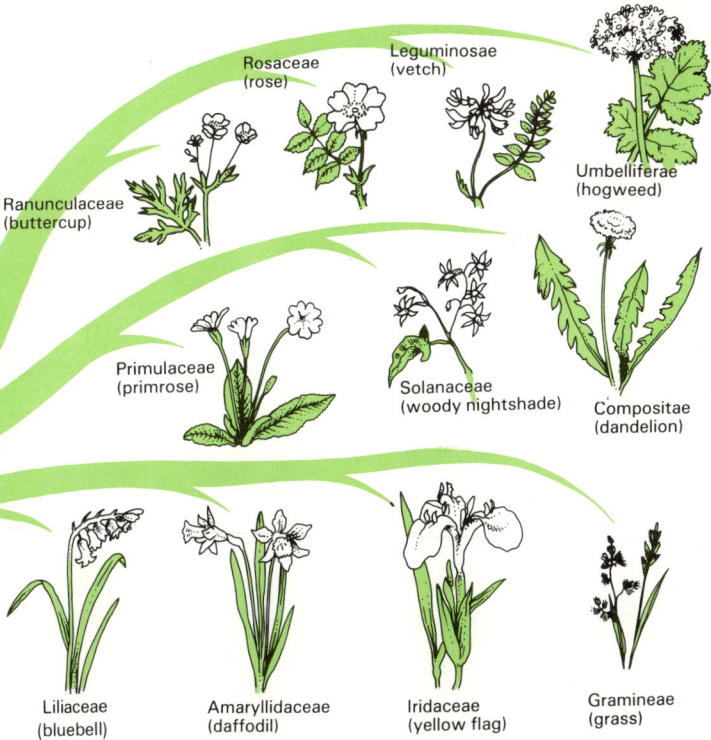

Ranunculaceae
(buttercup)

Rosaceae
(rose)

Leguminosae
(vetch)

Umbelliferae
(hogweed)

Primulaceae
(primrose)

Solanaceae
(woody nightshade)

Compositae
(dandelion)

Liliaceae
(bluebell)

Amaryllidaceae
(daffodil)

Iridaceae
(yellow flag)

Gramineae
(grass)

Plane Any of several large trees of the north temperate zone. They have large leaves deeply cut into five or seven lobes, and rounded clusters of small red (female) and yellow (male) flowers. The fruits form brown globular clusters. The grey bark flakes to reveal yellow patches.

Plankton The free-floating organisms near the surface of seas or lakes. It includes many crustaceans, jellyfishes, the young stages of fishes, and the host of tiny organisms — plant and animal — on which they feed.

Plantain Any of several low-growing herbs of temperate regions. They have a rosette of leaves with long stems bearing dense spikes of small green or brown flowers. Plantain is also the name of a kind of banana.

Plantain-eater, *see* TURACO
Plant-hopper, *see* HOPPER

Plantigrade Walking on the soles of the feet. Human beings and bears are plantigrade animals.

Plant kingdom The chart on the following pages depicts the major groups of plants in the form of a family tree, showing the evolutionary relationships between the various groups.

Plant lice, *see* APHID

Plasma The fluid of blood. Blood plasma contains food materials, carbon dioxide and other waste products from the tissues, hormones from the various glands of the body, and the antibodies that have been produced to combat infections. It also contains the necessary substances for blood-clotting.

Plastid A tiny body of which large numbers are found in the protoplasm of plant cells. Some (leucoplasts) are colourless and are concerned with

177

starch storage. Others contain chlorophyll and are called chloroplasts.

Platelet Particles found in mammalian blood which are probably fragments of blood cells. They play a part in clotting the blood when blood vessels are damaged.

Platypus Believed by scientists to be a hoax when first discovered in 1796, the platypus is an egg-laying mammal with a duck-like bill, a furry body and webbed feet. It is a semi-aquatic nocturnal creature, living in burrows close to fresh water in eastern Australia.

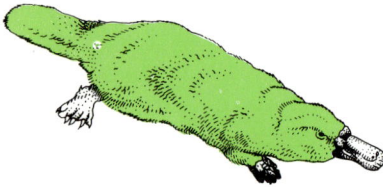

Duck-billed platypus

Plover A group of wading birds with long legs and short beaks which can often be seen running along the foreshore. They include among others, the ringed plovers, named for the black bands across the breast and face, and the golden plovers, distinguished by their black and gold spangled backs. Golden plovers are more likely to be seen on moorland.

Plum Any of several deciduous trees bearing short-stalked stone-fruits. Unlike peaches, the stone is generally quite smooth. Plums have long oval leaves and white flowers. They include the DAMSON, GREENGAGE and BLACKTHORN.

Plume moth A family of moths whose wings are normally divided into 'feathers' — the forewings into two or three parts and the hindwings into three or four parts. They have long slender bodies and long legs.

Plumule The first young shoot of a seedling.

Pochard A diving duck which rarely leaves the water except to roost and breed. Pochards live mainly in the northern hemisphere where they include the European pochard, the American canvasback and the tufted duck.

Pocket gopher A group of burrowing, hamster-like rodents living in North America. There are a number of species, up to 45 cm in length including the tail. They get their name from the fur-lined external pouches on their cheeks. These are used for carrying food.

Podsol A poor, sandy soil, from which the useful minerals have been washed to deeper levels by rain. Often supports heathland and coniferous forests.

Pogonophore A group of small worm-like marine animals. They are found in all the oceans, often at great depths. Pogonophores live in transparent tubes up to 5 times their own length. The name pogonophore means beard-bearer and refers to the tentacles at the front end which can number up to 200 in some species. These animals are placed in a phylum of their own — the Pogonophora.

Poison ivy Any of several bushes or creepers of North America that bring the skin up in blisters on contact. The bush types are often called poison oak. Poison ivies have leaves forming three leaflets, tiny white flowers and cream berries.

Polar bear One of the largest and most carnivorous of the bears, the male polar bear may reach a length of 2.75 metres. Females are a little smaller. The animal has a long head with small ears. It has powerful legs, with broad feet and hairy soles which help it to grip the ice and snow. Polar bears live along the southern edge of the Arctic pack ice. They eat seals and fish and will also take vegetable food when they come on land.

Polecat A member of the weasel family, the polecat is about 50 cm long,

Plume moth

including the tail. The upper surface has a dense, golden underfur with coarse, dark-brown guard hairs. The underside is blackish. The animal prefers wooded country, but lives more or less anywhere that it can find small animals to eat. Polecats are found throughôut most of Europe. In the British Isles they are confined to Wales and neighbouring parts of England at present.

Pollack A member of the cod family found in the eastern Atlantic from Norway to the Mediterranean. It is about 60 cm long, dark green above shading to white below. The sides are streaked with yellow.

Polar bear

Pollen The spores of seed plants that produce the male sex-cells.

Pollination The transfer of pollen grains from stamens to stigma: the first stage in the process whereby the male cells gain access to, and fertilize, the female egg-cells of seed-bearing plants. Many flowers are pollinated by insects, especially bees, as they move from flower to flower in search of pollen and nectar. Other flowering plants, particularly trees and grasses, are wind pollinated. They tend to produce large quantities of pollen since very little is liable to reach the female stigmas. When seeds are produced after the transfer of pollen from one flower to another (cross-pollination) the resulting plants are often stronger than if the pollen and egg-cell had both come from a single flower (self-pollination). Most flowers consequently have some way of avoiding self-pollination and

ensuring cross-pollination. Most flowers contain both stamens and carpels but a number of plants have flowers of one sex only. Some species (e.g. willow) even bear the male and female flowers on different plants. In these cases self-pollination is impossible. Where there are organs of both sexes in a flower, self-pollination is often avoided by the stamens ripening before or after the stigma is ready to receive pollen. Although cross-pollination is preferable, self-pollination is better than no pollination at all, and in many cases the stamens and stigma bend towards each other before the flower dies so that self-pollination may occur if cross-pollination has failed.

Polyanthus A perennial garden flower derived by a complex breeding from a cross between PRIMROSE and COWSLIP. The flowers range in colour from pale yellow to deep red or purple.

Polyp The fixed, tube-like form of the COELENTERATES.

Polypetalous flowers have free petals — i.e. not joined.

Pome A false fruit formed mainly from the receptacle which swells up around the seeds. The apple and pear are good examples.

Pomegranate A bush or small tree of western Asia, producing large golden fruit tinged with red. The leaves are lance-shaped, and the tubular red flowers form small clusters.

Pompadour fish A group of disc-shaped fishes up to 20 cm long, found in rivers of the Amazon basin. They are extremely colourful fishes, with vertical bands of blue or green and brown. The colours develop and change with age.

Pond skater A small bug with a flat narrow body and two pairs of long legs. These have hairy tips which enable the insect to rest on the water surface. The legs also 'row' the insect about. The front legs are short and are used for catching small insects that fall on to the water surface.

Pondweed Any of more than 100 small plants, mostly of temperate regions, that live in still fresh water. Roots and some leaves are under water, while other larger leaves float. They have small flowers.

Lombardy poplar

Poorwill The poorwill or whip-poor-will is an American nightjar about 20 cm long and mottled brown in colour. Like the other nightjars, the poorwill is nocturnal and catches insects on the wing. It is the only bird known to hibernate.

Poplar The name of 35 species of quick-growing trees related to WILLOW. They have oval or triangular leaves on flattened stalks that flutter in the slightest wind. They include ASPENS.

Poppy The name of more than 100 long-stalked flowers with long, cut leaves. The stems and fruits yield a milky juice, and the four-petalled blooms are bright and showy although short-lived. The juice of the unripe seed-capsules of the Asian opium poppy contains opium, a drug from which heroin, codeine and morphine are made.

Porbeagle A shark, normally about 2 metres long, living in the Atlantic Ocean. It is a plump fish, with a sharply pointed snout and a large dorsal fin. It feeds mainly on other fishes.

Porcupine The name given to a group of rodents with sharp quills on the back which can be erected as a very effective means of defence. The quills are also rubbed together when a porcupine is excited which produces a loud rustling sound. The Old World porcupines have long quills which reach up to 45 cm in length in the common porcupine of southern Europe and northern Africa. The New World porcupines are forest-dwelling creatures with shorter spines.

Porcupine fish This fish is about 30 cm long and has numerous spines lying flat on the body. When alarmed, the fish swallows water and puffs itself up. The spines stand out from the body then and the fish is very difficult for any enemy to swallow. Porcupine fishes live in tropical seas and feed on corals and shellfish.

Porpoise A group of small whales, none of which exceeds two metres in length. They have blunt snouts and broad rounded flippers. These features distinguish them from the other major group of small whales, the dolphins. There are several species, generally

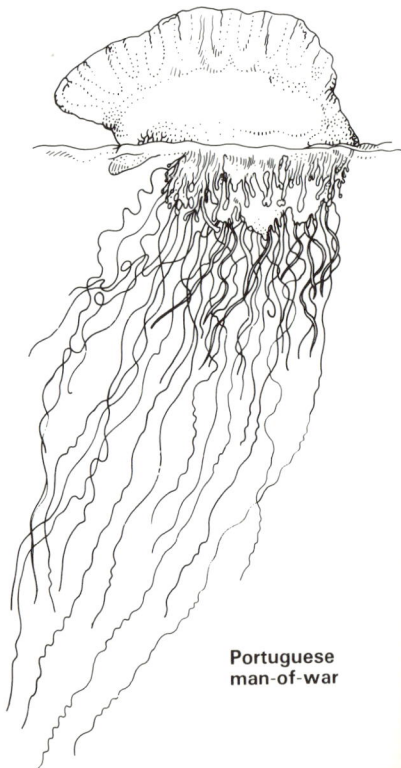

Portuguese man-of-war

found in coastal waters. They swim in the surface waters, catching herrings and other small fishes.

Port Jackson shark A small shark, about one metre long, found in the seas off eastern Australia. It has a heavy head with prominent nostrils either side of a tooth-filled mouth. The two dorsal fins each have a stout spine in front.

Portuguese man-of-war A marine animal — actually a colony of minute animals — related to the jellyfishes and sea anemones. There is a gas-filled float at the surface, and below it are numerous long tentacles with stinging cells. These contain a powerful venom. The man-of-war lives in the warmer seas of the world and is often brought to British waters by the Gulf Stream. It feeds on fishes.

Potato A tuberous rooted plant of South American origin, widely grown as food. It is related to nightshades and tomatoes. It has rough compound leaves, small white, purple or yellow star-shaped flowers, and small tomato-like poisonous fruits.

Potoo Also known as wood nightjars, the potoos live in South and Central America. They have mottled brown plumage, large eyes and wide gaping mouths. Potoos roost during the day and consequently are rarely seen; they are mainly known by their mournful calls.

Potter wasp A wasp so named because the female builds a small flask-shaped nest attached to a plant from grains of sand cemented with saliva. She lays one egg in the nest and stocks the nest with caterpillars paralyzed by her sting. She then seals the entrance and flies away to repeat the process. When the larva hatches it feeds on the caterpillars until it emerges from the nest as an adult.

Potto A slow-moving relative of the lemurs and lorises, up to 40 cm long. It lives in the forests of west and central Africa. It has thick brown fur and large round eyes, enabling it to see as it moves around at night. It grips the branches tightly and never lets go with more than one foot at a time.

Prairie chicken A plains-dwelling grouse of the central United States.

Once widespread over the grasslands of North America, from southern Canada to Texas and eastwards to the Atlantic, the range of the prairie chicken is now very restricted.

Prairie dog A hamster-like ground squirrel which has a barking call. It is about 30 cm long and yellowish-brown or grey in colour. The five species live on the plains of North America. They are burrowing creatures and they live in 'towns' of many family groups, each group having several burrows in a small area. Also known as prairie marmots.

Prairie hare, see JACK RABBIT

Prairie wolf, see COYOTE

Pratincole A group of birds related to coursers found in Africa, Asia and Australia. About 20 cm long, they look like large swallows in flight, with pointed wings and forked tails. They are sometimes called swallow plovers. Pratincoles live in open country, often near water, and feed on insects.

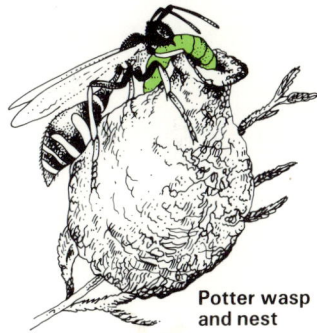

Potter wasp and nest

Prawn A name originally given to a shrimp-like crustacean with a toothed 'sword' on its head about 6 cm long and found around the coasts of Europe. Today it is applied to many long-bodied crustaceans. The largest live in tropical seas and reach 30 cm in length.

Premolar One of the cheek teeth of mammals.

Primate A member of the highest order of mammals which includes man, monkeys, apes, lemurs, tarsiers

Pronghorn

and tree-shrews. The limbs of a primate are in most cases adapted for life in the trees. All five fingers and toes are well developed and the thumb and big toe can be moved to the other digits in order to grasp objects. The female usually bears only one young at a time.

Primrose The name of about 500 species of perennial herbs with deeply-veined leaves. They have funnel-shaped flowers which appear in spring. The common primrose is yellow, but many other shades exist.

Privet The name of about 50 evergreen shrubs related to lilac. They have small leaves, strongly-scented white flowers, and tiny dark-skinned fruits. They are poisonous.

Proboscis The trunk of elephants; also the coiled 'tongue' of butterflies and moths.

Proboscis monkey A brick-red LANGUR with a swollen, drooping nose hanging down over the mouth. Proboscis monkeys live in the forests of Borneo, often near rivers.

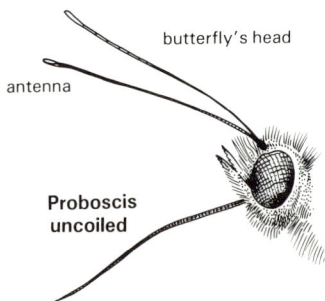

butterfly's head

antenna

Proboscis uncoiled

Processionary moth A group of moths found in many parts of the Old World whose caterpillars are famed for the processions they form. Some, such as the pine processionary, form a single line of 200 or more caterpillars. Others, such as the oak processionary, form shorter wedge-shaped columns.

Prolegs The stumpy hindlegs of caterpillars. They disappear completely in the adult insect.

Pronghorn An antelope-like creature living on the grasslands of western North America. It stands about one metre at the shoulder. The upper parts are reddish-brown, with a black mane; the underparts and rump are white. Both sexes bear horns covered with a sheath of hair. The horns have a short prong at the front — hence the name — and they are shed each year.

Propagation Increase in numbers by any means, though the gardener normally restricts the term to increase by vegetative means — i.e. other than sowing seed. One of the commonest methods is the taking of cuttings. This involves cutting a piece of stem (with some plants it can also be done with roots or even leaves) and putting it into moist but well-aerated soil until it takes root and begins to grow as a new plant. Layering is a form of propagation allied to cutting and is used especially for carnations. A shoot is partly cut and the cut region is pegged into the soil until it takes root. The rooted portion is then detached from the parent plant. Budding and grafting involves joining a bud or a shoot (the scion) of the desired variety on to the stem of another variety (the stock).

Protein A group of complex organic (carbon-containing) compounds which are an essential part of all living organisms. They are built up from amino-acids and contain carbon, hydrogen, oxygen and sometimes sulphur and phosphorous. They are incorporated into many hormones and enzymes. Green plants are able to build up their own proteins from raw materials, whereas animals rely on their food as a source of protein.

Prothallus The stage in a fern life-cycle that bears sex-cells and alternates with the spore-bearing fern plant.

Puma

Protista A term sometimes applied to single-celled organisms.

Protoplasm The jelly-like substance within a cell which is the living matter of the cell.

Protozoan A member of the phylum Protozoa which contains single-celled animals widely found in nature. Most are minute and live in the sea and fresh water, though some live in damp soil and others live inside larger animals. Many have short hairs (cilia) or long whip-like hairs (flagella) which beat to move the animals from place to place or to draw a current of water containing food and oxygen over them. Some are naked masses of protoplasm while others have elaborately patterned shells of chalk or silica.

Pseudopodium A temporary projection of protoplasm put out by a cell for the purpose of moving or engulfing food and other particles.

Ptarmigan This game bird is distinguished from its grouse relatives by its white wings and underparts. It ranges right across the northern parts of the world. The birds have a brownish mottled colour in spring and summer. In autumn they become greyish, and in winter they are white except for the black tail and the black eye-patches of the male.

Pteridophyte Any of a large division of the plant kingdom (Pteridophyta) which includes ferns, horsetails, clubmosses and related plants.

Pudu The smallest American deer, standing 40 cm high at the shoulder. Pudu deer range the lower slopes of the Andes from Colombia to southern Chile.

Puff adder A group of poisonous African snakes that have an especially loud hiss. They include the brightly-coloured Gaboon viper of the tropical forests and the more sombre common puff adder of deserts and grasslands.

Puffball Any of several kinds of fungi whose fruiting bodies are globular, and puff out spores when touched.

Puffer The name given to a group of fishes all of which can blow themselves up to twice their normal size. Most live in shallow tropical seas and some grow to 30 cm or more in length.

Puffin A black and white seabird with a triangular bill of red, blue and yellow. These colours develop only during the breeding season. The birds eat small fishes and they swim under the water scooping up the fishes as they go. Puffins live around the North Atlantic and breed on cliffs.

Puma Also called a cougar or mountain lion, the puma is a large cat look-

Ptarmigan

winter   summer

ing rather like a lioness, though rather greyer in colour. The body is generally about one metre long. The animal is found throughout the Americas, especially in the mountains and plains. It also lives in forests. It feeds mainly on deer, but may also take domestic animals.

Pumpkin A trailing plant producing large, spherical fruit. It has large prickly leaves. It is closely related to the SQUASH. There are several species.

Pupa The resting stage, often called the chrysalis, through which insects pass while they change from larvae into adults.

Pupil The opening in the iris of the eye which allows light to fall on the retina.

Purse sponge A vase-shaped sponge about 5 cm long found around the coasts of Europe from the Arctic to the Mediterranean. Purse sponges live on rocky shores between high- and low-tide levels.

Puss moth Named for the fluffy hairs which cover its body, the puss moth is a mottled grey moth found in many parts of Europe and Asia. The caterpillars are 5 cm long. When disturbed they rear up to present a highly coloured and patterned front end looking much like a large face. They also wave slender 'tails' to discourage enemies.

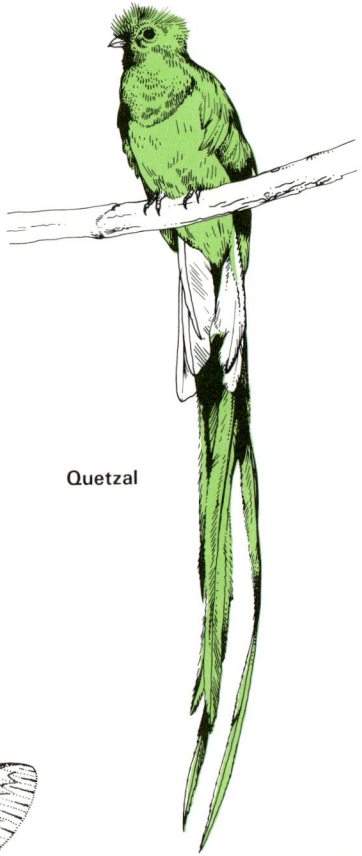

Quetzal

Puss moth

adult

caterpillar

Pyrethrum The name of several species of CHRYSANTHEMUM. An insecticide is made from the dried flowers, which are cultivated on a large scale in East Africa and some other areas.

Python A group of large constricting snakes, of which the Asian reticulated python is the largest. It reaches a length of almost 10 metres. Pythons live in various habitats, but the larger ones prefer to be near water. Many climb trees. They eat many kinds of mammals and birds.

Quack-grass, *see* COUCH-GRASS

Quail Game birds resembling small partridges. There are nearly 100 species, living all over the world. They do not fly much, and keep within a metre or so of the ground when disturbed. They keep to open ground, feeding on seeds and insects.

Quelea The quelea is a sparrow-like bird with a stout red beak. It lives in vast flocks on the grasslands and savannas of Africa and does immense damage to grain crops. Whole fields of grain may be eaten.

Quetzal A bird of Central America famed for its plumage, especially the long iridescent green 'train' formed by the tail coverts. The quetzal featured in Aztec mythology, and its plumes could only be worn by nobles. In more recent times the 'quetzal' has become the Guatemalan unit of currency.

Quickthorn, *see* HAWTHORN

Quince A small fruit tree closely related to the APPLE. The tree is many-branched with large pink or white flowers. The yellow pear-shaped fruit is sour-tasting.

Quinine, *see* CINCHONA

Quokka A small rat-like wallaby. It is less than one metre long, including the tail, and it is greyish-brown with a reddish tinge around the front part. The quokka lives in the swamps and brush of south-western Australia, feeding on grasses and other plants at night.

Radial symmetry

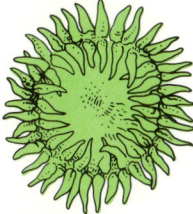

Sea urchin

Starfish

R

Rabbit The rabbit can be distinguished from the hare by its shorter ears and limbs. Rabbits live in extensive burrows or warrens with several emergency exits. Their diet consists of a wide variety of plant food including grass and many important crops. During the summer the female may have a number of litters at intervals of 5–6 weeks. Rabbits seem to have spread through Europe from Spain. They have also been introduced into other continents. In Australia rabbits have destroyed vast areas of grassland. There are over 50 races of domesticated rabbits.

Raccoon Carnivorous mammals related to pandas. There are a number of species, all living in the Americas. They are between 60 cm and 1 metre long including the tail. The fur is generally greyish with dark marks on the face and tail. Raccoons live mainly in the forests, especially around streams.

Raccoon fox, *see* CACOMISTLE

Raceme A type of inflorescence with stalked flowers on a main axis (e.g. bluebell).

Racer A group of non-venomous snakes with slender bodies and large heads found in the warmer, drier parts of the world. The largest is the black racer, over one metre long.

Radial symmetry means the arrangement of limbs and other organs around a central point. An example is the starfish. Radial symmetry is characteristic of sedentary animals.

Radicle The root of a seed-plant embryo.

Radiolarian A group of single-celled marine animals remarkable for the intricacy of their tiny skeletons. The largest is only 50 mm across, and most are far smaller. Radiolarians occur in vast numbers and are important planktonic animals. Large areas of the ocean floor, particularly in the tropics, are covered with ooze formed from radiolarian skeletons.

Radish

Radish A small annual herb related to MUSTARD. It has small, rough leaves and white flowers. It produces large spherical or cylindrical red or white spicy roots, which are eaten in salads.

Radius A bone of the forearm on the thumb side.

Radula The scraping tongue of gasteropod molluscs.

Rafflesia Any of six Malaysian plants which have no stems or leaves. They are parasitic on climbing shrubs and consist mainly of slender threads which spread through the host plant. Huge flowers spring from the base of the host plant from which the threads have drawn all their nourishment. The flowers, measuring up to 90 cm across, smell like rotting meat and are pollinated by small flies.

Ragweed, *see* HOGWEED

Ragworm A marine relative of the earthworm living in shallow coastal waters all over the world. There are many species. Each segment bears broad lobes and tufts of bristles and, unlike the earthworm, the ragworm has a fairly distinct head with eyes and jaws. Ragworms range from 2 cm to 1 metre in length. Most of them are carnivorous.

Ragwort A perennial herb of the daisy family, native to Eurasia. It has stems up to 1.2 metres tall, ragged leaves, and bright yellow flowers.

Rail Long-legged, short-tailed birds, most of which live in and around water. Coots and moorhens are rails. They are generally rather shy birds and run for cover when disturbed. They do not fly very strongly. Many species have loud and rather eerie calls.

Rainbow finch, *see* GOULDIAN FINCH

Rainbow wrasse, *see* WRASSE

Rain frog A rounded brownish frog about 3 cm long. It lives in the dry regions of southern Africa and sleeps (aestivates) for much of the year. It comes out after the rains and feeds for a couple of months or so on worms and termites. The eggs are laid under the ground and the tadpole stage is passed entirely in the egg. When the egg hatches, a tiny frog emerges.

Rape A leafy plant of the cabbage family, similar to SWEDE. Its quick-growing leaves are used as a pasture crop and to make winter feed for livestock. The seedlings are often sold with cress as mustard; whole seeds are also a valuable source of oil.

Raspberry A fruit-bearing bush with tall prickly stems. New stems grow every year and die after bearing fruit in the second year. The small red fruits are actually a collection of tiny DRUPES.

Rat A long-tailed rodent, particularly the brown rat (sometimes called the Norway rat) and the black rat (sometimes called the ship rat). The brown rat, which, despite its alternative name of Norway rat, originated in Central Asia, is the larger of the two, with a head and body measuring 20–25 cm and a tail equally long. It is much the commoner pest in both towns and countryside and has spread throughout the world. Being omnivorous it is now a serious agricultural pest in many places. The black rat is slightly smaller in the body but its tail is as long as that of the brown rat. The black rat originated in South-East Asia but, like the brown rat, has spread throughout the world as a stowaway on board ships. It was the black rat that was responsible for the spread of bubonic plague across Europe in the Middle Ages. Outside the tropics it is now confined to urban areas, particularly docks. The so-called water rat is actually a vole.

Ratel

Ratel A relative of the wolverine, the ratel is a black and white badger-like animal measuring about 90 cm including a short tail. It is found over most of Africa south of the Sahara as well as in

India. Like badgers, ratels are omnivorous, but they are particularly fond of honey and their alternative name is the honey badger.

Ratite A term sometimes used for the large flightless birds, such as the ostrich, kiwi, emu, etc.

Rattan palm The name of several climbing palms, natives of the Far East. The reedy stems of these palms climb over other trees to a length of 150 metres or more. Rattan stems are used for basket-work.

Rattlesnake The rattlesnake is so named because of the horny rings on the tail. These are the remains of cast-off skins and they rattle when the snake is disturbed. There are various species in North America, some of them dangerously poisonous. One species may exceed 2.5 metres in length. They all give birth to live young.

Raven The largest member of the crow family, about 60 cm long and completely black. It ranges over much of Eurasia and North America. Similar species live in Africa and Australia. Most ravens live in mountainous and coastal regions, and they feed largely on carrion and small mammals. They nest in trees or on rocky ledges.

Ray A group of cartilaginous fishes with flat bodies and broad pectoral fins. They include the electric ray and skate among many others.

Razorbill A black and white seabird related to the puffins and guillemots. It is a plump bird about 40 cm long, with a white stripe crossing the black beak. Razorbills are confined to the North Atlantic and spend much of the year out at sea. They come inshore during the winter and breed on cliff ledges in the spring.

Razor shell A group of bivalve molluscs with long narrow shells. Most grow to about 15 cm in length. Their resemblance to the old-fashioned 'cut-throat' razor has earned them their European name. An equally similar resemblance to a penknife is responsible for the North American name of jack-knife clam. Razor shells live on the shore and burrow into the sand.

Receptacle The top of the flower-stalk where the petals and other flower parts arise.

Red deer

Receptor A sense organ: one that responds to a stimulus by producing nerve impulses which are carried to the central nervous system by sensory nerves. Different receptors respond to different stimuli, such as light, heat, pressure and smell, and enable an organism to monitor its external and internal environment. Receptors may be single cells or they may be complicated structures such as the eye.

Red admiral A familiar migratory butterfly with a vivid colour scheme of black, scarlet and white. The forewings have a fine tracery of blue near the outer margins. The red admiral is widely distributed over the northern hemisphere. In Europe it winters in Mediterranean regions and flies north in early summer to lay its eggs on nettle leaves.

Red beetroot, *see* BEET

Red bird, *see* CARDINAL

Red deer The European red deer stag stands about 1.3 metres at the shoulder and carries antlers up to one metre long. The hind is considerably smaller. The sleek reddish-brown summer coat is replaced in winter by a thicker, greyish brown coat. A stag's antlers are shed every year between January and April. The new antlers are covered with soft skin until they are fully grown and during this time the stag is said to

Reindeer

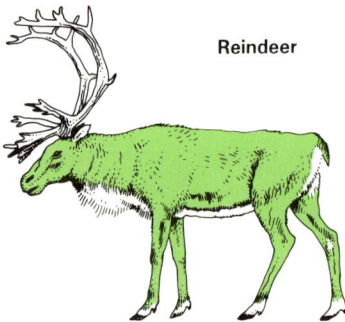

be 'in velvet'. Red deer are browsing animals whose natural habitat is forest. In Britain, where they are largely forced to graze on moorlands, the animals are smaller.

Red eft, *see* NEWT

Red fox The European red fox and the closely related North American form are both about 75 cm long in the body with a bushy 40-cm tail. The female or vixen is slightly smaller. The thick fur is sandy to reddish-brown above and whitish on the underparts. The legendary cunning of the fox is a measure of its adaptability: the animal prefers wooded regions but is frequently found in urban areas, scavenging in refuse bins. The natural diet of a fox ranges from ground-nesting birds and small mammals to insects and vegetable matter.

Red mullet, *see* MULLET

Red panda, *see* CAT-BEAR

Redpoll A small sparrow-like finch found in northern parts of Europe, Asia and North America. It has a forked tail, a crimson forehead and a black chin. Redpolls feed on seeds and small insects.

Red river hog, *see* BUSHPIG

Red squirrel, *see* SQUIRREL

Redstart A shy relative of the robin found in Europe and Asia. The common redstart has an orange breast and rump in the male, the female having buff underparts. The black redstart is almost wholly black in the male, the female being dark brown. Both have brick-red tail feathers. In North America the name redstart is applied to a group of warblers.

Red-tailed monkey, *see* GUENON

Redwood A large coniferous tree of the western United States. Redwoods grow to more than 90 metres tall and may live for thousands of years. They are related to SEQUOIAS.

Reed Any of several tall, slender GRASSES, living in swamps or shallow ponds and streams.

Reedbuck A small graceful antelope found on the grasslands of Africa south of the Sahara. Standing about 80 cm at the shoulder, the male carries ringed horns hooked forwards at the tips.

Reed frog There are many species of reed frog, all living in Africa. They are only about 2.5 cm long, but many species are beautifully coloured and marked. The markings often vary from individual to individual and depend to some extent on the background.

Reed-mace, *see* BULRUSH

Reeve, *see* RUFF

Reflex action An immediate response to a stimulus not controlled by the will (e.g. withdrawal after touching a hot object). The stimulus sends an impulse directly through nerve fibres to the spinal cord. From there it is relayed to appropriate muscles for bringing about the correct response.

Regeneration The ability to replace parts of the body that have been lost through accident. All animals and plants are capable of it to some extent but, as a general rule, the more highly evolved an animal is, the less are its powers of regeneration. Human beings, for example, are able to regenerate skin and bone tissue to mend wounds and fractures. But some animals (e.g. lizards and crabs) are able to regenerate new limbs while others, such as flatworms, can grow complete bodies — including the head — from a small portion of the original body.

Reindeer The reindeer of northern Europe and Asia is a semi-domesticated race of the CARIBOU.

The coast redwood is the tallest tree in the world, with one specimen in California reported as being 112 metres high.

189

**The human
respiratory
system**

trachea

bronchus

ribs

lung

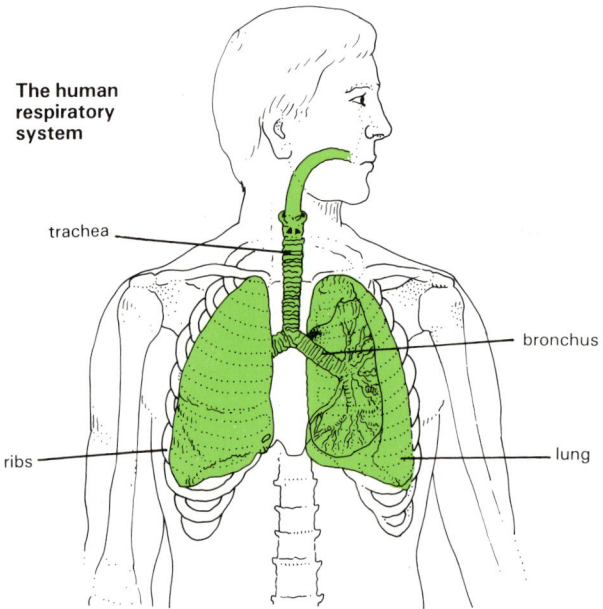

Remora A fish in which the dorsal fin has been converted into a large oval sucker. The remora attaches itself to the undersides of other fishes. Sharks commonly carry remoras in the warmer seas of the world. The remora does no harm to the shark and often moves away to feed on passing small fishes. Remoras may reach one metre in length but most are much smaller.

Renal Concerning the kidneys.

Reproduction A term which simply means producing more of the same kind. Nothing lives for ever, and if plants and animals could not reproduce, their species would soon die out. There are two main methods of reproduction — sexual and asexual.

Sexual reproduction always involves the joining together of two special cells called GAMETES to form the new generation. The joining together is called fertilization. In some simple plants and animals the gametes are all alike, but most organisms produce two very different kinds of gametes, known as male and female. The male gamete is usually very small and active, and it generally seeks out the larger and often immobile female gamete. Both kinds of gametes may be produced by one organism, especially among the plants, and these organisms are said to be bisexual or HERMAPHRODITE. Most animals, however, are unisexual: each individual is either male or female and produces either male or female gametes. In most animals the male gamete (sperm) is brought into contact with the female gamete (ovum) through the act of mating. Plants cannot mate, however, and most of them rely on wind or animals to carry their male gametes (in the pollen grains) to the female organs (see POLLINATION). The fertilized ovum, whether plant or animal, is called a zygote and, by repeated cell division, produces an embryo which gradually develops into the new plant or animal.

Asexual reproduction involves no fusion of cells and in its simplest form, found among the unicellular plants and animals, it involves nothing more than the splitting of the body into two parts. Many higher plants produce special detachable buds which grow into new individuals, while the production of new plants from RUNNERS, as in the strawberry, is another form of asexual reproduction. Hydra and many other coelenterates can reproduce in a similar way by budding or branching: new animals grow as branches on the parent and often break away to lead their own lives. Parthenogenesis is a special form of asexual reproduction in which the females lay fertile eggs or give birth to young without mating. It is particularly common in insects, including aphids and stick insects. Unlike sexual reproduction, in which the characteristics of two parents are combined in the offspring, asexual reproduction produces offspring which are genetically identical to the parent. It does not produce variation, which is so important for EVOLUTION. (See also ALTERNATION OF GENERATIONS; HEREDITY.)

Reptile A cold-blooded vertebrate belonging to the class Reptilia and suited for life on land. The skin is dry and covered in scales. Reptiles have lungs and show an advance on amphibians in that they lay shelled eggs which hatch on land. They are widely distributed but are most abundant in warmer regions. Examples include CROCODILES, SNAKES, TURTLES and LIZARDS.

Respiration One of the essential life processes, respiration is carried out by all living things in order to get energy. The normal respiratory process can be summed up by the following equation:

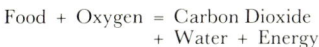

Food + Oxygen = Carbon Dioxide
+ Water + Energy

Glucose sugar is the usual food material used in the reaction, which takes place in all living cells. Although respiration is strictly just the biochemical reaction taking place inside the cell, the meaning of the word is commonly extended to cover the various ways in which animals and plants obtain oxygen for use in this reaction. In this sense it means the same as breathing. Plants and simple animals get enough oxygen by simple diffusion from the surrounding air or water, and the waste carbon dioxide is eliminated in the same way. More complex animals, however, need special respiratory or breathing organs, such as GILLS and LUNGS, to gather in the required amounts of oxygen. Most of them also need a transport system — the blood — to carry the oxygen around the body. Insects have a very different system of respiration, with minute tubes called tracheae carrying air from the surface to all parts of the body.

Respiratory movement Movement of parts of an animal which keeps either the water or the air from which the animal receives oxygen constantly fresh. Human beings renew the air in their lungs by movements of the chest and diaphragm.

Resurrection plant A small herb of Arabia, Egypt and Iran. When adult it loses its leaves, the branches curl into a ball, and it blows around until the rainy season, when the branches uncurl and new leaves appear.

Retina A light-sensitive layer found inside the eye-ball. It is developed only at the back of the eye. There are two distinct layers making the retina. The innermost layer is transparent and contains the light-sensitive structures (rods and cones); the outer layer is heavily pigmented. Nerve impulses caused by light hitting the retina pass to the brain along the optic nerve. The place where the optic nerve joins the back of the eye-ball has no retina. This is the blind spot.

Rhea This flightless ostrich-like bird is the largest bird in the New World. It stands about 1.5 metres high and, like the ostrich, it has an almost featherless head and neck. The long powerful legs have three toes, compared with only two in the ostrich. Rheas live on the grasslands of South America.

Rhesus monkey One of the sacred monkeys of India, the rhesus is about 45 cm long with a 25-cm tail. It ranges in colour from grey-green to yellow-brown on the back with a white belly. When frightened a rhesus becomes red in the face. It belongs to the group known as macaque monkeys. There are several species living in India and South-East Asia.

Rhinoceros

Rhinoceros

Rhinoceros A bulky thick-skinned mammal related to the horse. Both animals bear their weight on the central toe of each foot, although the rhino does have two extra toes on each foot. Rhinos bear one or two horns on the snout, composed of densely compacted fibres like hair. There are five species: two in Africa, one in India, and two more in South-East Asia. The last three are very rare. They range between 1 metre and 2 metres in height and the Indian rhino weighs up to 2.5 tonnes.

Rhinoceros beetle Several beetles, including some of the world's largest insects, have been given this name. Two horns projecting from the thorax and head give the rhinoceros beetles a ferocious appearance but they are harmless creatures. Their grubs feed mainly on decaying wood and other rotting vegetation. Rhinoceros beetles live mainly in the tropics.

Rhizome A fleshy horizontal underground stem.

Rhododendron Any of 900 species of trees and shrubs of Asia and North America. They are mostly evergreens, with long waxy leaves and twisted branches, and they produce large clusters of purple, white or red flowers. AZALEAS are deciduous relatives of the rhododendrons.

Rhubarb A perennial herb with huge, heart-shaped leaves. The red leaf stalks are edible, but the leaves themselves are poisonous. Rhubarb is native to Mongolia and Siberia.

Ribbon fish, *see* OARFISH

Ribbonworm A large group of long, often flattened creatures mostly living in the sea. They are usually less than 20 cm long but some species such as the bootlace worm can measure 10 metres or more. All have a sticky or barbed proboscis which they fire out to capture their prey.

Rice A cereal grass which grows best in warm, wet places. It is native to Asia, and is the main food of half the world's population.

Rickettsia A micro-organism similar to but smaller than a bacterium and behaving like a virus. It can cause diseases, such as typhus.

Rifleman A New Zealand wren about 10 cm long with a needle-like bill. It is an insect-eating bird which hops up tree-trunks probing crevices in the bark.

Right whale A group of whalebone whales which got their name because they were the 'right' whales to catch when men first started whaling. None of them is common today because they have been hunted for a long time. The Greenland right whale is one of the rarest of the whales.

Ringed seal One of the commonest seals, this species is mostly confined to Arctic waters, though there is an isolated population in the Baltic. The ringed seal measures about 1.5 metres in length. Its fur is generally pale grey with darker spots ringed with white.

Ringtail, *see* CACOMISTLE

RNA (Ribonucleic acid) A nucleic acid found in all living cells which plays an important part in the synthesis of proteins. Viruses carry their genetic instructions on RNA rather than on DNA as in all other organisms.

Roach A common freshwater fish of Europe and Asia reaching 45 cm in length. It is deep-bodied, greyish-green on the back and greenish-gold on the flanks. The fins are tinted orange to red. The rudd looks similar but is more heavily built. There is also a distinct difference in the position of the fins. In the roach the pelvic fin is immediately below the dorsal fin. In the rudd the pelvic fin is attached well forward of the dorsal fin.

Roadrunner An American member of the cuckoo family famed for its speedy running and its ability to catch snakes. Its wings are short and it rarely flies, but it has a long tail which helps to balance the bird as it runs. The roadrunner lives in the deserts of North and Central America.

Roan antelope Standing up to 1.5 metres at the shoulder, the roan antelope has large ears, an erect mane and curved horns up to 75 cm long. It is found on the African savannas.

Robber crab A land-living crab found on the islands and shores of the Indian Ocean and the western Pacific. Rob-

Rock rose

ber crabs climb trees, though for what reason is uncertain. They feed on carrion, fruit and the pulp of broken coconuts.

Robber-fly Also known as assassin-flies, robber-flies have needle-like mouthparts for piercing and sucking the blood of their prey. Some robber-flies look like horse-flies, others resemble wasps and bees. They feed on other insects ranging from small bugs to beetles and even bees, often catching them in mid-air.

Robin The name applied to many birds with a red breast in different parts of the world. The original robin red-breast is the European robin, a plump little bird with a conspicuous reddish-orange forehead, throat and breast in both sexes. It is a familiar garden bird, noted for its boldness towards human beings and its aggressiveness towards its own kind. The American robin is an equally familiar and favourite garden bird, more closely related to the fieldfare than the European robin. In Australia and New Zealand the name is applied to several species of fly-catcher.

Rockdove Also known as the rock pigeon, the rockdove is the ancestor of the domestic pigeon and the feral 'town' pigeon. Wild rockdoves live in Europe, Asia and North Africa, often in dry, rocky areas. Their chief food is seed. Pigeons were raised for food in ancient Egypt and during their long period of domestication many varieties have been bred.

Rock eel, *see* DOGFISH

Rock pigeon, *see* ROCKDOVE

Rock rose A trailing mat-forming or cushion-forming shrub of Eurasia and

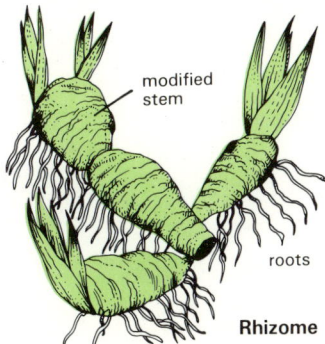

modified stem

roots

Rhizome

Rocky mountain goat

North Africa. It has oblong hairy leaves and short-lived yellow flowers, although there are red- and pink-flowered cultivated varieties. The rock rose is not a true rose.

Rock wallaby A group of wallabies found in rocky areas in most parts of Australia. Rock wallabies are extremely agile creatures, capable of climbing almost vertical rock faces. They also scale trees.

Rocky Mountain goat A goat-antelope found in the mountains of north-western North America. It has long white hair and slightly curving horns about 20 cm long. Males stand up to one metre at the shoulder. Rocky Mountain goats live in pairs or in small herds.

Rocky Mountain sheep, *see* BIGHORN

Rod (of eye), *see* EYE

Rodent A gnawing mammal belonging to the order Rodentia which includes rats, mice, squirrels, beavers, etc. Rodents have a pair of continuously-growing incisor teeth in each jaw. These teeth have enamel only on the front surface so that they have sharp cutting edges. The capybara is the largest rodent at the size of a small pig.

Roe deer Though widespread throughout Europe and Asia and increasingly common in many wooded places, roe deer are not often seen. They are adept at moving stealthily through the undergrowth and feed mainly at dusk and in the early morning. Standing 75 cm at the shoulder they are the smallest of the native European deer. In summer the thick coat is short and reddish-brown, in winter it is longer and grey-ish-brown. The antlers of the male, which are shed in November and December, are rarely longer than 20 cm and seldom more than three-pronged. Roe deer live in small family groups — not in herds like many other deer.

Roller Sturdy, jay-like birds with long curved beaks and long tails. Their plumage is largely blue, green and red. Most of the rollers live in Africa, but one species breeds as far north as Scandinavia. They get their name from their rolling courtship flight.

Rook

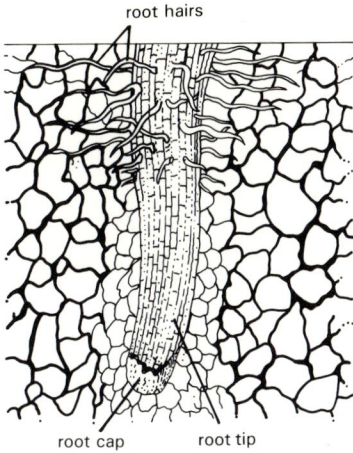

root hairs

root cap   root tip

Rook Though similar in size, shape and colour to the carrion crow, the rook can be distinguished by the grey patch of skin around the base of its beak. The plumage of a rook also has a bluish sheen compared with the greenish sheen of the crow and the feathers on its upper legs give the impression that the bird is wearing a pair of short baggy trousers. Rooks are gregarious birds and gather in huge flocks to build a colony or rookery of nests in tall trees, and to feed on a wide variety of plants and animals, ranging from grubs to grain.

Root The main functions of a root are to anchor the plant and to absorb and transport water and dissolved salts. Normally, roots grow downwards under the influence of gravity and away from light. This reaction enables a sprouting seed to get a hold on the soil. There are two basic patterns of root growth — the tap-root system (e.g. thistle) and the fibrous system found, for example, in grasses.

Root nodule A swelling found on the root of leguminous plants caused by the presence of certain bacteria. These bacteria are able to convert free nitrogen into a form which can be used by the plant.

Rorqual A group of whalebone whales including the blue whale, the fin

whale, the sei whale, Bryde's whale, and the minke whale or lesser rorqual. Bryde's whale is found in the warmer seas, but the others live in the polar seas and are the mainstay of today's whaling industry, although the blue whale is now extremely rare.

Rose The name of a family of 2000 plants, but especially of a number of shrubby species with beautiful flowers. They have leaves divided into leaflets, and most have prickly stems. Roses may be erect, climbing or rambling.

Rosemary An evergreen shrub of southern Europe, related to mint. It has small slender leaves, green on top and white below, and bluish flowers. The leaves are used as flavouring and yield an oil employed in perfumes.

Ross seal The smallest of the Antarctic seals, measuring about 2 metres in length, Ross seals are plump animals with large protruding eyes. They are seen only on the Antarctic pack-ice and little is known of their habits.

Rotifer A large group of microscopic animals found mainly in fresh water. The largest is just 2 mm across and most are less than 0.5 mm. Rotifers have a large variety of forms: they may be worm-like, spherical or flattened and soft-skinned or armoured, but they usually have cilia at the 'front' end which they wave to move through the water.

Roundworm, *see* NEMATODE

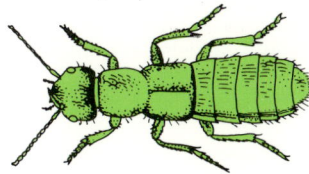

Rove beetle (devil's coach horse)

Rove beetle A group of beetles with long wings, but elaborately folded under wing covers. Most are very small, but some, such as the devil's coach-horse, may exceed 2.5 cm in length. Many are black or black and red. Most of them live in damp places, feeding on decaying matter and smaller insects.

Rowan, *see* MOUNTAIN ASH

Royal antelope The smallest of the antelopes, standing only 25 cm at the shoulder. Royal antelopes are found in West Africa.

Rubber tree Any of several trees and bushes yielding latex, but especially the hevea tree, or pará rubber tree of Brazil. It is a tall, slender evergreen, producing pale yellow flowers.

Ruddy duck, *see* STIFFTAIL

Rue A perennial small shrub of Europe. It has much-divided bluish-green leaves, with a strong smell and bitter taste. The flowers are small and yellow. Rue is used in medicine.

Ruff A wading bird in which the male has an ornate ruff or collar of feathers around the neck. The male is about 30 cm long; the female, or reeve, is a little shorter. Ruffs breed in northern Europe and Asia, inhabiting damp meadows and marshes. Out of the breeding season they migrate as far as southern Africa and Australia.

Ruminant A cud-chewing animal, such as deer, antelopes, sheep, cattle and giraffes. All belong to the order Artiodactyla and are herbivorous. Food is swallowed without chewing and is passed into the rumen — the first chamber of a complicated stomach. From there it is later regurgitated into the mouth for chewing — a process known as 'chewing the cud'.

Runner A horizontal shoot that takes root at the tip and forms a new plant.

Rush Any of a number of plants with hollow or pith-filled stems. They have flat or stem-like leaves and clusters of very small brownish flowers. They generally grow in marshy ground in the cooler parts of the world. Rushes are used for basket work and chair bottoms.

Rust One of the many important parasitic fungi of the class Basidiomycetes. Rusts cause serious diseases of cereals and other plants.

Rye A cereal grass used extensively as a food crop. It has its grains in long ears with whiskers, the so-called 'beard'. It is grown mainly in the cooler regions.

S

Sable A marten found in the forests of Siberia. It has a dark brown or blue-black fur tipped with white. Sables measure about 50 cm in length and have a short bushy tail. They have long been hunted for their fur.

Sable antelope Standing 1.25 metres at the shoulder and armed with a pair of long scimitar-shaped horns, this is one of the few antelopes that will defend itself vigorously if attacked. Even lions seem loth to attack a sable antelope. These antelopes are found on the African savannas south of the equator.

Sage A small shrubby plant of the mint family. It has woolly greyish-green leaves and produces blue, purple or white flowers. Leaves and stems are used for flavouring.

Sagebrush A shrub of the western American plains. Its wedge-shaped leaves smell like sage when crushed. It has tiny yellow or white flowers, and can grow up to 3.5 metres tall.

Sago palm Either of two East Indian palm trees. The trees flower when 15 years old, but are cut down just before to get the starchy pith out of the stems. From this the sago used as a food is prepared.

Saiga An antelope with a noticeably swollen snout. It has features of both sheep and gazelles. It is about 75 cm tall and it has a woolly coat which is buff in summer and white in winter.

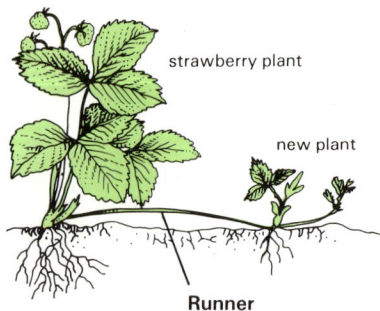

strawberry plant

new plant

Runner

Salamander

The male has lyre-shaped horns. The saiga lives on the cold steppes of Central Asia.

Sailfish A highly streamlined, fast-swimming fish up to 3 metres in length found in all tropical waters. It may be the fastest of all fishes and has been credited with speeds of more than 100 km/h. The upper jaw is extended into a sword-like beak and the first dorsal fin forms a large 'sail' when fully raised.

St John's wort Any of 300 species of herbs and low shrubby plants, also called hypericums. They have long narrow leaves which often bear lots of translucent glands. The flowers are yellow with numerous protruding stamens.

Salad burnet, *see* CALCICOLE

Salamander The salamanders belong to the group of amphibians with tails — the Caudata. They are found in most parts of the northern hemisphere in brooks and ponds and in moist spots on land. Most salamanders are under 15 cm in length but the giant salamander of Japan reaches 1.8 metres from the snout to the tip of its tail.

Saliva A fluid secreted into the mouth by the salivary glands. Saliva lubricates the mouth and food before swallowing. Digestive enzymes may be present (e.g. ptyalin in man).

Salmon A highly prized game and food fish found in the northern parts of Europe, Asia and North America. Specimens weighing 35 kg have been recorded, though 10 kg is a more usual figure. Adult salmon live in the sea but they return to the river in which they were born to spawn, leaping rapids and small waterfalls on their journey upstream. When the young salmon hatch they make their way gradually downstream, taking between one and eight years to do so depending upon the temperature of the water: the lower the temperature the slower the growth of the fish and the longer the time taken. During the various stages of its growth a salmon is known successively as an alevin (fry), fingerling, parr and smelt.

Salsify A biennial herb with long narrow leaves resembling GOATSBEARD. It has long tapering edible roots tasting like oysters, and is sometimes called the oyster-plant. Salsify was formerly cultivated as a vegetable and harvested before the purple dandelion-like flowers appeared. It is now established on grasslands and waste-ground.

Saltbush Any of 1400 species of herbs and shrubs. They resist dry conditions and can grow in salty ground.

Saltwort An annual seashore plant with a striped, many-branched stem. It has spine-tipped leaves and tiny greenish flowers and thrives in salty ground. There are several similar species.

Samphire Any of several species of fleshy perennial herbs growing among rocks and on sand-dunes on the coast. The wedge-shaped leaves are broken into leaflets, and the flowers are small and yellow or white.

Sand dollar This creature, also known as a cake urchin or sea biscuit, is a rather flattened form of sea urchin.

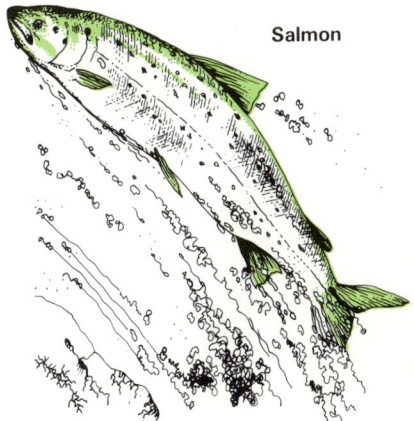

Salmon

197

The shell is purple or black and is up to 10 cm across. There are several species, found mainly in the warmer seas, especially in the Pacific.

Sandalwood Any of several trees growing in southern Asia and Australia. They have oval leaves, and the timber is very hard and sweet-smelling.

Sand gaper, *see* CLAM

Sand-hopper, *see* AMPHIPOD

Sand rat, *see* GERBIL

Sapodilla A tropical American tree with leathery, evergreen leaves. It produces sweet, apple-shaped fruit. Chicle, the main ingredient of chewing gum, comes from its latex.

Saprophyte An organism that obtains its food from dead organic material. Many bacteria, the majority of fungi, and some flowering plants live in this way.

Sapsucker A North American woodpecker, so called because of its habit of drilling holes in trees and sucking the oozing sap. There are two species, both about 20 cm long and brightly coloured. The birds also catch the insects that are attracted to the sap.

Purple saxifrage

Sapwood The soft outer region of wood found just beneath the bark. Made up of fresh xylem tissue, it is the active region of water conduction in woody plants. The heartwood in the centre of the wood is very compact and the xylem vessels are crushed and unable to carry water.

Sardine, *see* PILCHARD

Sausage fly, *see* ARMY ANT

Savoy cabbage, *see* CABBAGE

Sawfish Large fishes, occasionally up to 6 metres in length, with a long snout

Scarab beetle

flattened into a blade and a row of strong teeth sticking out of each side. This blade may account for as much as one-third of the fish's length. Sawfishes are found in warm coastal waters. They feed on molluscs and crustaceans, using the saw to search in the mud.

Sawfly The sawflies are related to bees and wasps, but do not have the narrow 'wasp waist'. Their name is derived from the fact that they have a saw-like edge to the ovipositor with which the female cuts slits in plants to lay her eggs. Some sawflies are called wood wasps because they lay their eggs in wood.

Saxifrage A family of more than 1000 species of herbs, shrubs and trees, and especially a group of 300 rock plants. They grow mostly in Arctic and mountainous regions. Most have low rosettes of leaves and bear small flowers on long stalks. LONDON PRIDE is a typical species.

Scabious A group of small annual and perennial herbs with lacy, flat-topped flowers rather like those of the Compositae family, except for the stamens which stand out from each flower. The flowers are often blue or mauve. The plants grow between 45 and 90 cm tall.

Scale insect Small bugs distantly related to greenfly. They are among the most serious of crop pests. The males normally have one pair of wings, but the females are wingless and they merely sit on plants and suck sap. They are flattened creatures, never more than 2.5 cm long and usually much smaller. Many are protected by horny scales which make them look like tiny limpets.

Scallop Marine bivalve molluscs in which one valve of the shell is normally flat and the other convex. The valves are ridged, and they bear 'ears' near the hinge. Scallops are among the few bivalves that can swim. They do it by opening and closing the shell quickly and shooting water out.

Scaly anteater, *see* PANGOLIN

Scarab Stoutly built, blackish beetles with brilliant metallic sheens in many species. Most scarabs feed on the dung of mammals and are called dung beetles. Many bury the dung and feed on it underground. They also lay their eggs on it. Some species, such as the sacred scarab of Egypt, roll balls of dung about until they find a suitable place to bury them.

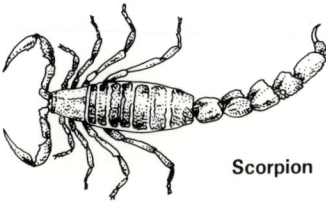

Scorpion

Scilla The name of about 80 species of Old World bulbous plants in the lily family. They have long strap-like leaves and blue or purple flowers. Several are cultivated and some garden forms have white flowers.

Scimitar oryx, *see* ORYX

Scion, *see* PROPAGATION

Sclera, *see* EYE

Sclerotic The tough coat of the vertebrate eye-ball.

Scorpion A group of arachnids renowned for the sting carried arched over the body. The venom of some species can be fatal to humans. Scorpions measure up to 20 cm in length, though most species are considerably smaller. They are found in warmer parts of the world, particularly in deserts.

Scorpionfish A group of fishes whose fins are normally broken up into numerous ribbons and spines, the latter often being very poisonous. Members

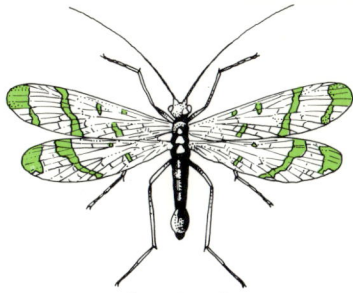

Scorpion fly

of the group include the highly poisonous stonefish and the Norway haddock. Although the fishes have poisonous spines, their flesh is quite pleasant to eat.

Scorpion fly A small group of insects so named because in some species the tip of the male's abdomen is carried arched over the back in a scorpion-like fashion. The head has a pronounced 'beak' with biting jaws at the end and most species have two pairs of wings. Scorpion flies are mainly scavenging insects.

Scrub A plant community dominated by shrubs and bushes.

Scrub-bird The name given to two wren-like Australian birds which live in the dense scrub and are rarely seen. Nests are built on the ground and consist of grass and twigs lined with 'papier mache' made from chewed-up wood.

Sea anemone A group of simple marine animals related to jellyfishes. The body is soft and jelly-like and

Beadlet sea anemone

199

Seahorse

bears numerous tentacles. The tentacles are covered with stinging cells which are used to catch fishes and other small animals. Sea anemones live attached to the rocks and shrink into small round 'blobs' when the tide goes out.

Sea bass, *see* GROUPER

Sea cow, *see* SIRENIAN

Sea cucumber Relatives of the starfishes and the sea urchins, sea cucumbers have sausage-shaped bodies equipped with rows of tube-feet and a mouth at the front end fringed with tentacles. Most sea cucumbers are less than 30 cm long. They live on the seabed, usually at depths of less than 200 metres.

Sea fir Also known as hydroids, sea firs are colonies of tiny polyps related to *Hydra* and to corals. They resemble delicate seaweeds in appearance. They are found in all seas, most living between the tide marks or in shallow waters. There is a free swimming larval form which settles on any solid object where it grows into a stalked colony of polyps through BUDDING.

Sea gooseberry, *see* COMB JELLY

Sea hare A group of molluscs with similarities to sea slugs and sea snails, so named because of a pair of broad tentacles which look rather like the ears of a hare. They have a soft body and a thin transparent shell. The sea hares of temperate waters are usually less than 10 cm long. They feed on seaweed.

Sea heath A perennial herb of Eurasia and Africa, living in salt marshes. It creeps along the ground and sends up wiry branches with tiny oblong leaves. It has small pink flowers.

Sea holly A coastal plant of Europe, with large bluish-green prickly leaves. It has dense heads of blue thistle-like flowers. Its roots used to be candied to produce a sweet called eringoes. Although thistle-like, the sea holly belongs to the carrot family.

Seahorse The seahorse is about 20 cm long and has a large head with a tubular snout, a rotund body and a very long tapering tail. It is almost entirely covered with bony bumps which give it the appearance of a wood carving. Despite its name and appearance, the seahorse is a fish found in most coastal waters. There are several species.

Seal A group of flesh-eating mammals specialized for life in the sea. They have a tapering body like that of a fish and broad limbs in the form of flippers. There is a thick layer of fat (blubber) beneath the skin. The group may be divided into the true seals, the eared seals and the walruses. The true seals have no outer ears to protect the ear passage, but the ear-holes close when the seal submerges. The front limbs are weak and the animals can only wriggle on land. The eared seals have small external ears and their front limbs are stronger: they can move about more easily on land. Walruses can also move about on land but they lack external ears and the canine teeth of the upper jaw are extended into tusks.

Sea lavender A herb of temperate regions, growing mainly in coastal salt marshes. The bright leaves spring directly from the rootstock, and it has branches of tiny pink or mauve flowers.

Sea lily Although very plant-like in shape, the sea lily is an animal related to the starfishes and sea urchins. In most species the body has five arms and is carried at the top of a long slender stalk which anchors it to the seabed. Sea lilies are found in deep water in most of the world's oceans. They feed on particles of decaying matter that float down from above.

Sea lion A group of eared seals related to the fur seals. The most familiar species is the Californian sea lion, which is often seen in circuses. This is the smallest of the five species, with a maximum length of about 2 metres.

Sea mat, *see* MOSS ANIMAL

Sea mouse A stout marine bristle-worm, related to the ragworms. It looks and moves rather like a mouse, hence the name. The animals live in shallow seas and are usually about 18 cm long.

Sea otter A marine relative of the common otter found in isolated colonies along the coasts of the northern Pacific, particularly off Alaska and California. Growing up to 1.25 metres in length including a 30-cm tail, the sea otter has a dense, glossy coat, brownish-black on the back and creamy white on the head, throat and chest. Hunted almost to extinction for its soft fur, this animal is now carefully protected.

Sea pen Related to the sea anemones and sea firs, these animals often do look like old-fashioned quill pens stuck in the mud of the seabed. They mostly have a central shaft like the quill of a feather with branches bearing polyps. Like their relatives, sea pens have a free-swimming larval form which settles on the seabed and turns into a polyp. The animal then grows by BUDDING. Sea pens are found in all the oceans. Some grow between the tide marks, but the majority live in deeper water. The largest measure 2 metres in length, though most are much smaller.

Sea perch, *see* GROUPER

Sea slug A group of marine snails which do not have shells. They are rather flattened creatures, but the upper surface bears numerous flaps and filaments which act as gills. Many of the species are brightly coloured.

Sea snake A group of aquatic snakes, normally about 1.5 metres long, found in tropical waters, mainly in the Indo-Pacific region. They often live near sandy shores and river mouths. All sea snakes are venomous with cobra-like fangs at the front of the mouth.

Sea spider A small-bodied, long-legged marine creature which resembles a spider, though the only connection between the two is that both are arthropods. Sea spiders are found in all the oceans, from the seashore down to 4000 metres. They feed on other animals such as coral polyps and sea anemones. Most measure just a few centimetres across, though deep-sea forms have been recorded measuring 60 cm across the legs.

Sea squirt Marine animals which, in the adult state, live permanently fixed to the seabed or to rocks. Many live in clusters or colonies. Sea squirts look like little bags of jelly, but they are in fact related to vertebrates. The young sea squirts are tadpole-like creatures with a primitive backbone.

Sea lion

Sea urchin

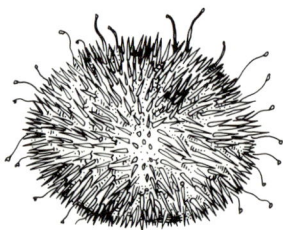

Sea urchin

Sea urchin A small round animal with a spiny shell which lives around rocky coasts. The shell or test of a sea urchin is made up of hundreds of interlocking chalky plates. The sea urchin moves about on small water-filled tube-feet which stick out through pores in the test. There are many species including the purple sea urchin, the heart urchin and helmet urchin.

Sea wasp, *see* JELLYFISH

Seaweed Algae living around the shore, and sometimes far out to sea. They may be green, brown or red. Seaweeds have no leaves, although many of them are leaf-shaped. The main part of the plant is known as the frond and it is normally attached to a rock by a saucer-shaped organ called a holdfast. Green seaweeds get their colour from the substance chlorophyll which colours the leaves of land plants.

They are often found in rock pools and other places where the water is less salty. Brown and red seaweeds also contain chlorophyll, but the characteristic green colour is masked by other pigments. Brown seaweeds include some of the largest species, such as kelps (also known as oarweeds). Kelps form dense underwater forests with fronds occasionally over 50 metres long in some Pacific species. Wracks are also large brown seaweeds. Red seaweeds are generally smaller and some can be found in deeper waters, down to about 20 metres. They include some edible seaweeds, such as purple laver which is boiled to make laver bread.

Secretary bird A bird of prey whose name comes from the crest of black-tipped feathers which hang down behind the head in the way 18th century clerks carried their quill pens stuck into their wigs. Adult male secretary birds stand over 1 metre at the shoulder and have a wingspan of 2 metres. The plumage is grey, with black on the wings and legs. They feed on various small animals and also kill snakes. Secretary birds are found in Africa, south of the Sahara.

Sedge Any of a group of perennial tufted herbs, generally found in damp habitats in the temperate zone. They have long, grass-like leaves and the stems carry spikes of male and female flowers. Some species grow in wood-

Seaweeds

Bladder wrack

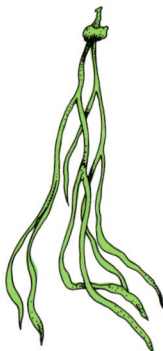

Sea thong

Sea lettuce

land or on sand. The stems are always triangular in section and the leaves form a triangular pattern when viewed from above.

Sedum, *see* STONECROP

Seed The reproductive body of flowering plants, conifers and a few other plants such as seed ferns. The seed is formed from the fertilized ovule and contains an embryo and a food reserve. The embryo consists of a radicle (root), plumule (stem), and seed leaves (cotyledons).

Segmentation The division of the body of an animal into a number of similar parts or segments. It is especially marked in earthworms where each visible segment contains structures which are repeated, with slight variations, in all other segments.

Sei whale, *see* RORQUAL

Self-pollination Pollination of a stigma by pollen from the same flower.

Semicircular canals Organs of balance in the vertebrate ear.

Senna A large group of shrubs and herbs with yellow flowers and leaves broken up into leaflets. Sennas are also called cassias, and are related to peas. A purgative is made from the leaves and seed pods of some species.

Sensitive plant Any of several tropical species of MIMOSA whose leaflets fold up when the plant is touched. The whole plant may collapse if repeatedly touched.

Sensory nerve A nerve which carries impulses from a sense organ to the central nervous system.

Sepal Leaf-like parts outside the petals of a flower which form the calyx.

Sequoia The largest tree in the world, although not the tallest. It grows only in the mountains of California. Sequoias are evergreen conifers, with small scale-like needles. The largest, the General Sherman Tree, is probably 3500 years old. Sequoias are related to the REDWOOD, and are sometimes called Big Trees or Wellingtonias.

Serow A goat-antelope standing almost one metre at the shoulder with a long mane and strongly ridged horns about 20 cm long. Serows are found in the Himalayas, where they live up to 3000 metres, southwards to

Giant sequoia

Malaya and Sumatra. Closely related forms are found in Tibet, China and Japan.

Serrate (of a leaf) With a toothed edge, like a saw.

Serval A long-legged member of the cat family found in Africa south of the Sahara. It stands about 45 cm at the shoulder. The coat of the serval is yellowish-brown with a pattern of bold black spots and stripes.

Service tree Either of two Eurasian and North African trees. The wild service tree has maple-like leaves and acid fruit; the true service tree has compound ash-like leaves and sweet fruit. They are related to the MOUNTAIN ASH.

Sesame An Indian herb producing oily seeds. It has pointed oval leaves, and white flowers spotted with purple, red or yellow. The oil is used in cooking and food preparation.

Sessile Without a stalk.

Seville orange, *see* ORANGE

Sex-cell A cell which has half the normal number of chromosomes (haploid) and which combines with another sex-cell during fertilization to produce a ZYGOTE. Sex-cells are also known as gametes. The female sex-cell of seed-producing plants and animals is called the ovum, the male sex-cell of seed-producing plants is the pollen and of vertebrates, the sperm.

Sex chromosome One of the pair of CHROMOSOMES that determine the sex of humans and most other animals. They are known as the X and Y chromosomes and, although they pair up during the formation of sex-cells at MEIOSIS, they are not exactly alike: the X is often much larger than the Y. Among the mammals the female body cells contain two X chromosomes, while male body cells contain one X and one Y. When egg cells or ova are formed in the female's body each one receives an X chromosome. When sperms are formed in the male's body, however, each receives either an X or a Y. If the ovum is fertilized by an X-bearing sperm a female offspring will result (with two X chromosomes in the cells), but if the ovum is fertilized by a Y-bearing chromosome a male offspring will result. Because X- and Y-bearing sperms are produced in equal quantities, there are more or less equal numbers of males and females in the offspring. Most other animals have

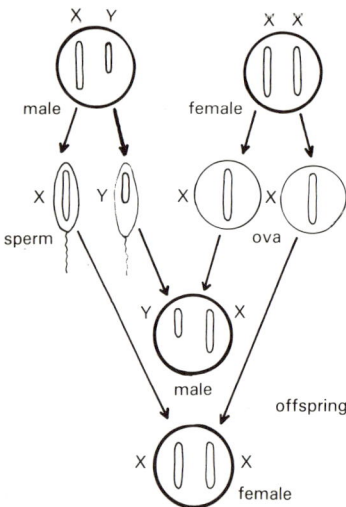

Sex determination

exactly the same system for determining sex, although birds and snakes and some insects have reversed it so that the male has two X chromosomes and the female has an X and a Y. Some plants, such as the hop, produce plants of different sexes and the mechanism is just like that of the mammals.

Sexton beetle, see BURYING BEETLE

Shad A group of herring-like fishes found in many coastal waters, rivers and lakes in the northern hemisphere. In European waters there are two very similar species, the allis shad and the twaite shad. In North America one of the best known species is the alewife.

Shag A sea-bird which is very similar to the cormorant but lacks the white chin and is somewhat smaller in size.

Shallot A form of ONION, producing clusters of small bulbs instead of a single large one.

Shamrock A plant with leaves consisting of three leaflets, the national emblem of Eire. There is, however, no definite plant of this name. Wood sorrel, yellow trefoil and white clover have all been claimed as the true shamrock.

Shark A large group of cartilaginous fishes that lack the scales of bony fishes. Instead, the tough leathery skin has tooth-like structures embedded in it. Sharks also lack the swim-badder of bony fishes and must keep swimming to stay afloat. Most sharks are marine fishes and range in size from the dogfish about one metre long to the whaleshark, the largest of all fishes, reaching 15 metres in length and weighing up to 15 tonnes. Sharks have several rows of teeth and eat animal food of all kinds. Many sharks are harmless to man including the basking shark and huge whaleshark.

Shearwater A group of petrels related to the albatross and fulmar. They are medium sized sea-birds with dull plumage, usually black or brown above and whitish underneath. Outside the breeding season shearwaters live far out at sea, and are found in many parts of the world excepting the polar seas.

Sheep A group of mountain-dwelling ungulates with spirally twisted horns. Wild sheep live in the northern hemisphere at heights up to 5000 metres. They include among others the MOU-FLON of Corsica and Sardinia, the ARGALI of central Asia and the BAR-BARY SHEEP of North Africa. Domestic sheep have probably been derived from various wild species. There are many varieties, some bred for their wool, others for their meat.

Sheetweb spider The familiar cobwebs found in houses and outbuildings are the work of house-spiders. Similar cobwebs are built amongst vegetation by many other spiders. All are loosely termed sheetweb spiders. Most of these spiders build a small tunnel beside the web in which they await their prey. The threads of a sheetweb are not sticky like the threads of an orb-web; when an insect blunders into a sheetweb the spider must seize its prey rapidly before it can escape.

Shelduck A brightly coloured, rather goose-like duck common in estuaries and on coasts. It is almost 60 cm long and the plumage is white with bold patterns of chestnut and black. The shelduck ranges from the British Isles, across Europe and Asia to China.

Shell A term loosely applied to any hard external skeleton but properly restricted to the hard covering of molluscs — whether a single valve as in snails or two valves as in mussels.

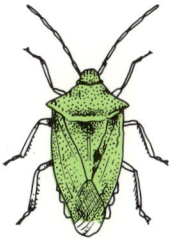

Shieldbug

Shieldbug A large group of sap-sucking plant bugs with flattened shield-shaped bodies. They somewhat resemble beetles in appearance and have a pair of hard forewings protecting the hindwings used for flying. Most shieldbugs are found in warmer countries and some are serious crop

Shoebill

pests. Many eject a foul-smelling liquid when disturbed and shieldbugs are consequently known by the alternative name of stinkbugs, especially in North America.

Shipworm A worm-like marine bivalve mollusc which feeds largely on wood and tunnels into underwater timbers. As the shipworm grows in size, so the tunnel increases in diameter. In the days of wooden ships the shipworm was a major pest since it is capable of reducing solid timber to a crumbling labyrinth of tunnels separated only by thin walls of wood.

Shoebill Named for its heavy broad beak, the shoebill is a stork found in marshy areas of central Africa. It stands about 1 metre high and has a wingspan of 2.5 metres or more. The plumage is grey with a greenish gloss. The shoebill's diet consists largely of frogs and small fishes.

Shoreweed A dwarf submerged plant, living in European ponds. It has grass-like leaves, and often forms an underwater turf. Its flowers are on stalks, exposed when the water is low.

Shoveller A dabbling duck with a spoon-shaped bill that frequents ponds and marshes. It breeds over much of temperate Asia, Europe and North America, wintering farther south.

Shrew The shrews are the smallest of the mammals. They are widely distributed in the northern hemisphere. The common shrew of Europe is about 10 cm long, including 4 cm for the tail. It is a mouse-like animal, but it has a much more pointed snout, full of sharp red-tipped teeth. Shrews feed mainly on insects and worms. They have to feed every two or three hours.

Shrike A group of small birds of prey found in all continents except South America. Shrikes perch in trees or hover over hedges in search of prey. Shrikes often build up a 'larder' by impaling their victims on thorns or sharp twigs, a habit which has earned them the name of butcherbirds.

Shrimp A small marine crustacean resembling a tiny lobster without claws. Edible shrimps grow to about 7 cm long. They are normally grey in colour, the flesh turning pink when cooked. Prawns differ from shrimps in having a toothed beak or rostrum projecting from the head.

Shrub A low woody plant, typically branching at or near the base and not developing much of a trunk.

Sidewinder A rattlesnake living in the deserts of western North America. It gets its name from its peculiar method of moving along the sand. It throws its body into curves, but only two points are in contact with the ground at any one time. The only tracks left in the sand are unconnected parallel grooves. Sidewinders are up to 60 cm long.

Sika deer Resembling a small red deer, the sika deer is native to Manchuria and Japan. It has been introduced into parks in other countries and now lives in the wild in a number of areas.

Silkworm The larva of a species of moth native to China which feeds on mulberry leaves and produces a coccoon of silk that is suitable for use in woven fabrics.

Silverfish A bristletail named for its bright silvery scales. This wingless insect is a common pest of kitchens and food cupboards.

Sinus A space within certain bones of the face, connecting with the nasal cav-

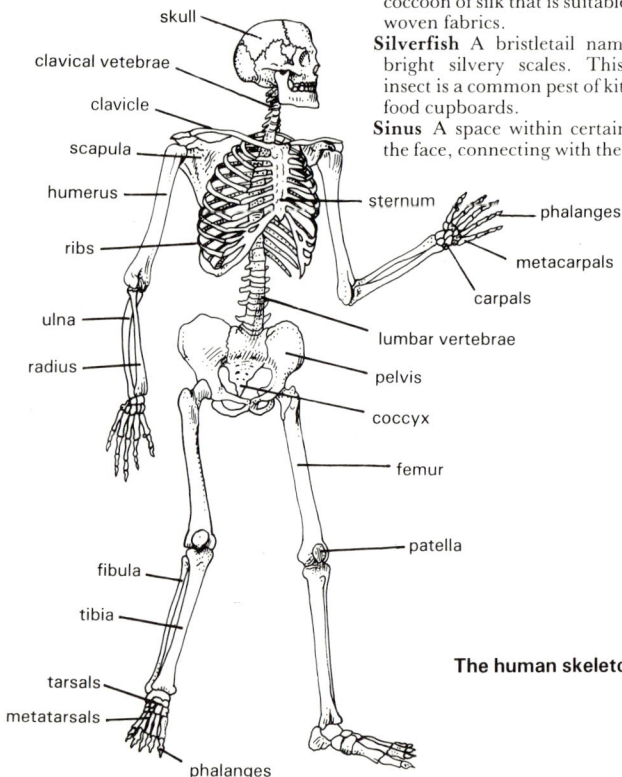

skull

clavical vetebrae

clavicle

scapula

humerus

ribs

ulna

radius

fibula

tibia

tarsals

metatarsals

phalanges

sternum

phalanges

metacarpals

carpals

lumbar vertebrae

pelvis

coccyx

femur

patella

The human skeleton

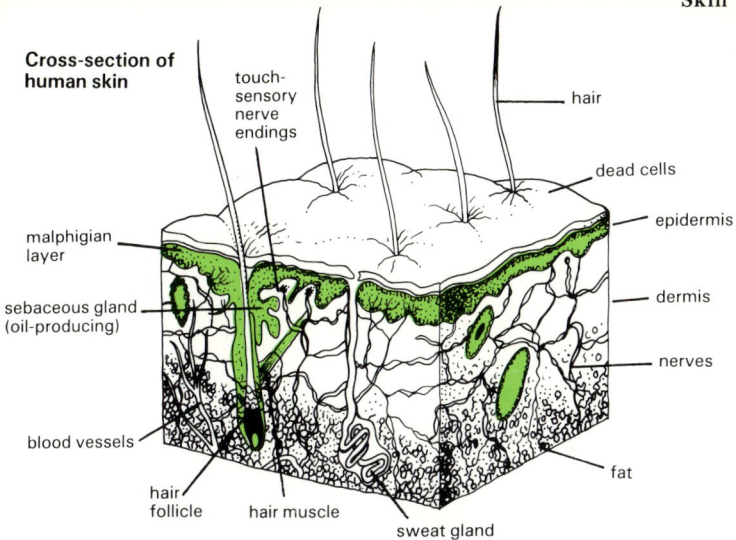

Cross-section of human skin

touch-sensory nerve endings

hair

dead cells

epidermis

dermis

nerves

malphigian layer

sebaceous gland (oil-producing)

blood vessels

hair follicle

hair muscle

fat

sweat gland

ity. Also an expanded vein of a type found especially in shark-like fishes.

Siren A small group of North American amphibians related to newts and salamanders. Adult sirens retain some of the features of young amphibians: their bodies are eel-like with no back legs and only weak front ones; and there are feathery external gills behind the head. Sirens range from 20 cm to 75 cm in length. They live in shallow streams and marshes.

Sirenian Also known as sea cows, sirenians form an order of aquatic mammals (the Sirenia) which contains the MANATEES and the DUGONGS.

Sisal, *see* AGAVE

Skate Cartilaginous fishes, related to sharks and rays. There are several species and they are all flattened from top to bottom. The body is diamond-shaped, with a slender spiny tail. Skates reach a length of one metre or more. They live in most temperate and tropical seas, especially in the northern hemisphere.

Skeleton A feature of most animals that gives shape and rigidity to the body, protects soft parts, and provides anchorage for the muscles. The skeleton is normally quite hard and may be inside (endoskeleton) or outside the body (exoskeleton). Exoskeletons include the shells of most molluscs and the hard coverings of crabs and other arthropods. Endoskeletons are found mainly in vertebrate animals and are normally composed of a series of bones, each able to move against its neighbour to allow movement of the body.

Skimmer The three species of skimmer are relatives of the gulls and terns, but they have long beaks with the lower half much longer than the upper. Skimmers are up to 50 cm long, with stout bodies and long wings. They fly low over water and scoop up fish with their beaks. The black skimmer lives around the coasts of the Americas. The other two species live in Africa and Asia.

Skin This is the outer covering of the body. In plants and many animals it is nothing more than a single layer of cells, often covered with a waxy or horny material produced by the cells themselves. It acts as a barrier, keeping dirt and germs out and keeping the body fluids in. It also helps to give the plant or animal its shape if it has a hard covering. In the more advanced animals the skin is very complicated and has many jobs. Hair, feathers, and nails are all outgrowths of the skin, and so are the scales of fishes and reptiles.

The mammalian skin has two main regions — the epidermis on the outside and the dermis underneath it. Between the two is the very important malpighian layer, which contains brown pigments — more in brown-skinned people than in white-skinned people, although extra pigments are built up when a person becomes sun-tanned. The cells of the malpighian layer are continually dividing and the outer ones produced in this way are added to the epidermis. But the latter does not keep getting thicker, because the outer cells are all dead and they are always flaking off. Scurf or dandruff is simply dead skin.

The dermis contains lots of nerves which are sensitive to touch and to temperature: they help an animal to feel things. The dermis also contains layers of fat and lots of tiny blood vessels, and it plays an important role in regulating the body temperature in mammals and birds. If the body temperature drops for any reason, the blood vessels near the surface close up. Little blood flows near the surface, and so little heat is lost. When a person goes pale it is because there is little blood flowing through the skin. Goose pimples, caused by the contraction of tiny muscles attached to the hairs on the skin, are not very important in

humans, but certainly are in other mammals. When an animal is cold, the muscles in the skin cause the hairs to stand on end, thus trapping a thicker layer of air around the body and helping it to keep warm.

If a person's body temperature rises, after a run or other strenuous exercise, the blood vessels in the dermis open up and let lots of blood flow near the surface. This makes the person look red, but also allows the extra heat to escape. Sweating is another important method of temperature control. When the body temperature is high, the sweat glands in the skin open up and pour water onto the surface. As this water evaporates, it cools the body. In really hot weather people need more to drink because they lose so much water through the skin in order to keep their temperature down.

Skink A large group of lizards found in many parts of the world. They are mainly burrowing creatures and have very small legs or none at all. The largest is about 60 cm long, but most species are much smaller.

Skipper A worldwide family of generally small butterflies. All have a broad head and body and many are sombrely coloured, particularly on the upper surface of the wings. Skippers normally make short, darting flights.

Skua Sea-birds related to gulls. There are four species, and they normally feed by attacking other sea-birds and forcing them to give up the fish they are carrying. The beak is hooked and the toes have sharp claws. The great skua breeds on the North Atlantic coasts and also around Antarctica. The other skuas nest around the Arctic.

Skunk This animal, related to the weasel and badger, is famed for its ability to squirt foul-smelling fluid at its enemies. There are several species, all with black and white shaggy fur. They live in the Americas.

Slipper animalcule, *see* PARAMECIUM

Sloe, *see* BLACKTHORN

Sloth South American mammals related to ant-eaters and armadillos. They spend most of their lives hanging upside-down in the trees by means of

Skunk

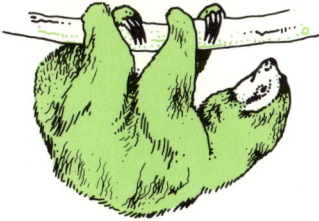

Sloth

their huge curved claws. The hair runs from the belly to the back, so that water runs off easily. Sloths grow to between 60 cm and 90 cm long. They move slowly and feed on leaves and fruit.

Sloth bear Also known as the Indian bear, the sloth bear has long shaggy hair and a long baboon-like muzzle. It can grow up to 2 metres long including the short tail. Sloth bears are creatures of uncertain temper living in the low-land jungles of India and Ceylon.

Slow-worm A legless lizard found over much of Europe and western Asia. The body is shiny brown with a black line down the back of the female. Adults are generally about 30 cm long, although they often lose their tails and appear much shorter. Slow-worms eat slugs and insects.

Slug A group of gasteropod molluscs closely related to the snails but mostly lacking a shell. They all live in damp places and are most active at night. They eat a variety of foods including cultivated plants and can be serious garden pests.

Smelt Small silvery fishes found only in the northern hemisphere. Smelt live in large shoals close to the shore and in brackish estuaries. The European smelt, up to 30 cm long, is one of the larger species.

Smooth newt, *see* NEWT

Smut A parasitic fungus of the class Basidiomycetes. Smuts commonly cause diseases of cereals, producing masses of sooty black spores on the leaves or flowers.

Snail The name commonly used for any gasteropod mollusc with a spiral shell. Snails include freshwater, marine and land-dwelling forms. Snails move by means of contractions of a muscular foot, leaving a tell-tale trail of slime behind them. In most land snails the head bears two pairs of tentacles with the eyes at the tip of the larger pair. Other snails have one pair of tentacles. Marine snails and some freshwater forms breathe with gills which lie in a cavity between the mantle and the body. In land snails the cavity has been converted into a lung and the gills have disappeared. Many freshwater snails are also air-breathing. Most snails feed on vegetable matter, but many marine species are carnivorous and have a rasp-like tongue or radula.

Snake A legless reptile belonging to the same order as the lizards (Squamata). Snakes are believed to have descended from lizards which took to living underground and gradually lost their limbs. Though they lack legs snakes can move easily over the ground. Some have broad scales on the belly which can be lifted and dug into the ground. Others throw their bodies into loops from side to side and glide forwards by pushing backwards against the ground. Snakes feed on other animals. The constricting snakes, such as the anaconda and python, squeeze their prey in their coils until it suffocates. Venomous snakes have hollow teeth called fangs through which they inject poison into their victim. Snakes are able to swallow prey wider than their own bodies because the lower jaw is separate from the skull, allowing the mouth to open very wide, and the ribs are unjoined on the underside, allowing the body to expand. Most snakes lay eggs with leathery shells, but some give birth to live young.

Snake-bird, *see* DARTER

Water snail

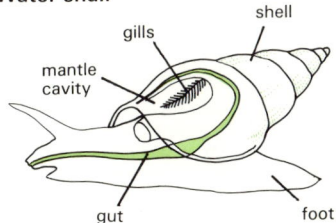

shell
gills
mantle cavity
gut
foot

Snake fly

Snowy owl

Snake fly A small group of insects so named because they can raise the head above the body like a snake about to strike. The female has a long ovipositor with which she lays her eggs in the crevices of tree bark.

Snake-necked turtle A group of long-necked turtles which withdraw the head into the shell sideways. They are found in South America, Australia and New Guinea. The Australian snake-necked turtle has the longest neck, more than two-thirds the length of the 35-cm shell.

Snakeroot The popular name for several plants whose roots look like snakes.

Snake's head, *see* FRITILLARY

Snapdragon A perennial European herb with tubular flowers that close with two lips. The lips open when the sides of the flower are gently squeezed or when the lower lip is weighed down by a bumble-bee searching for nectar. There are several species, some of which are cultivated.

Snapper A large group of deep-bodied fishes named for the way they suddenly open their jaws and snap them closed when they have been landed. Snappers are found in all warm seas. They grow up to 60 cm and some are important food fishes.

Snapping turtle Freshwater turtles of Central and North America which can inflict a painful bite if disturbed. Largest is the alligator snapping turtle of the south-eastern United States, weighing up to 100 kg. Snappers are largely scavengers, but do take small animals on occasion.

Sneezewort A creeping perennial herb of Eurasia and North America. It has narrow, stalkless leaves with toothed edges, and clusters of composite white flowers on tall stems. Its smell makes some people sneeze. Bachelor's buttons is a cultivated form with yellow-centred flowers.

Snipe Short-legged wading birds with long straight beaks and mottled brown plumage. They live in marshes and other wet places and use their beaks to probe for worms and other creatures in the mud.

Snoek An important food fish, the snoek is widespread in the seas of the southern hemisphere. It grows to more than one metre in length and can inflict a painful bite. It has a very long fin on the back and several small fins along the tail. Snoek is a Dutch word for the freshwater pike.

Snowdrop A bulbous Eurasian perennial herb related to the DAFFODIL. It has two or three narrow green leaves, and a solitary white flower on a long stalk. The three outer petals are much longer than the three inner ones.

Snowflake A European bulbous perennial herb related to the DAFFODIL but with six petals joined to form a bell. It has longer and more numerous leaves, and up to six flowers on each stem. There are several species.

Snow leopard Also known as the ounce, the snow leopard is about the size of a Labrador dog. It has a long bushy tail and a thick greyish coat marked all over with black spots. Snow leopards are found in central Asia, particularly in the Himalayas. During the summer they range high up the mountain slopes, and even in the winter they rarely descend below 2000 metres. Snow leopards feed mainly on small mountain animals, though they will attack sheep and goats.

Snowshoe rabbit, *see* HARE

Snowy owl A large white owl of the treeless tundra of the northern hemisphere. Some males are completely white but the females have dark bars and spots. The feathers extend down the legs to the claws.

Soapwort A Eurasian perennial herb whose leaves produce a lather when bruised in water. It has creeping roots, lance-shaped leaves, and clusters of fragrant white or lilac blooms. Soapwort is related to the campions.

Sole Tongue-shaped flatfishes usually up to about 30 cm long. They lie on their left sides and are almost completely surrounded by a fringe formed by the fins. Most species live in shallow seas in the warmer parts of the world.

Solenodon Shrew-like mammals found only in the West Indies. They are about 30 cm long excluding the tail which is 25 cm long. The head is relatively large and bears a long snout. Solenodons have remarkably long front feet and claws.

Solifugid A group of arthropods related to the spiders with large and powerful jaws in relation to their size. The body rarely measures more than 5 cm in length but solifugids prey on a variety of small animals including scorpions. Solifugids are found in the deserts of Africa, Asia and North America.

Solitary flower One borne alone on an unbranched axis.

Solomon's seal Any of several perennial herbs of Eurasia and North America. It has thick creeping roots which bear seal-like scars. The leaves are large and elliptical, and the flowers are bell-shaped and a greenish-white, hanging from long arched stems.

Song-thrush, *see* THRUSH

Sorghum A cereal grass of Africa and Asia flourishing in warm climates. It

has large broad leaves, and may reach a height of 5 metres. Some varieties are grown as grain or for animal feed; others produce a syrup.

Sorus A group of spore-forming capsules on the underside of a fern frond.

Soybean An annual herb of the PEA family. It has hairy stems, leaves divided into three leaflets, and purple flower-sprays. The plant is used as animal feed, and the beans yield oil, meal and flour. A native of China, it is now cultivated in many parts of the world.

Spadefoot A group of toads in which there is a horny projection on the side of each hindfoot. This is used for digging burrows. Spadefoots are widely distributed in the northern hemisphere.

Spadix The thick, fleshy flowering spike of Arum and other lilies.

Spanish fly, *see* BLISTER BEETLE

Spanish lynx, *see* LYNX

Spanish moss An EPIPHYTE of tropical and subtropical American forests. It is not a moss, but a herb with long grey stems, narrow leaves and yellow flowers.

Sparrow The name given to a large number of birds. True sparrows are seed-eating birds belonging to the genus *Passer*, from which the entire order of perching birds or Passeriformes gets its name. The best known species is the house sparrow whose range, like that of the house mouse, is now totally dependent upon human habitation. It was originally a native of the African savanna.

Spadefoot

Sparrowhawk

Spermatophere A 'packet' of sperm. The male spider deposits a spermatophore and carries it to the female in a specially modified appendage called a pedipalp. Male newts deposit spermatophores that the females pick up.

Spermatophyte The name given to all seed-bearing plants including flowering plants, conifers and a few others.

Sperm whale This is the largest of the toothed whales, the males being about 20 metres long. The head is huge but the lower jaw is relatively small. It contains 50 or more conical teeth, each up to 20 cm long. Sperm whales are found in all the oceans, but are most common in the warmer seas. They live mainly on squids.

Sphagnum moss Any of 350 species of moss from which peat is formed. Sphagnum moss grows in bogs, forming dense, spongy green or red cushions. Its stems and leaves absorb and hold water.

Sphinx moth, *see* HAWKMOTH

Sparrowhawk This handsome bird of prey lives in Europe, Africa and Asia. It has short, rounded wings and long legs. The plumage is brown above and white with brown bars below. The bird lives mainly in woodland and feeds on small birds. It has a characteristic flight consisting of bursts of rapid wingbeats alternating with long glides.

Spathe A large bract protecting a SPADIX. The name is also applied to the membranous covering protecting the unopened flowers of daffodils and related plants.

Spear grass, *see* COUCH GRASS

Species The smallest normally-used category of CLASSIFICATION. The members of a species are generally very much alike and can all interbreed. They cannot, however, cross with other species and produce normal fertile offspring.

Spectacled bear Named from the whitish markings around the eyes, the spectacled bear is the only South American bear. It is 1.5 metres long and has a dark shaggy coat. The animal lives in the forests of the Andean foothills from Panama to Bolivia.

Speedwell Any of a group of annual or perennial herbs with small blue or white four-petalled flowers. Speedwells are creeping plants, with oval, often toothed leaves. They are sometimes called veronicas.

Sperm A male sex-cell.

Spectacled bear

Sperm whale

Spider A member of the class Arachnida which also includes scorpions, mites and harvestmen. Spiders have two parts to their bodies, four pairs of walking legs and poison fangs with which to paralyze their prey. Apart from a similar body plan, another feature all spiders have in common is the ability to produce strands of silk from their bodies. But not all spiders use the silk to spin webs, and those which do differ in their skill: webs range from bluish tangled masses of threads to delicate symmetrical designs. Insects which become entangled in a web are quickly paralyzed by the spider's poison fangs. Spiders which do not spin webs have other means of catching their prey. The CRAB SPIDER lurks on flowers, changing its colour to match its surroundings, ready to bite an unsuspecting prey. WOLF SPIDERS run down their prey with a surprising burst of speed. Despite the reputation of the TARANTULA its bite causes nothing more than a slight irritation in human beings. But some spiders are extremely venomous, such as the notorious BLACK WIDOW spider of the Americas.

Spider crab A group of crabs with long spidery legs. The largest, the giant crab of Japan, has legs spanning up to 3.75 metres, though the body measures less than 30 cm in length. Most spider crabs are much smaller, some spanning only a few centimetres. They are found in coastal waters in most parts of the world.

Spider monkey Spider monkeys are South American monkeys with slender bodies and sparse, rather wiry fur. They are up to about 60 cm long and have a tail up to 90 cm long. The tail is remarkably prehensile and is used as an extra hand. Spider monkeys appear to eat nothing but fruit.

Spike An inflorescence with stalkless flowers on a main axis.

Spinach An annual herb grown as a vegetable. It originated in western Asia. It has wide, succulent leaves. When mature it produces a stalk on which flowers appear, male and female usually on separate plants. There is also another plant commonly called spinach. This is spinach-beet, a form of beetroot with leaves like true spinach and succulent edible leaf stalks.

Spinal column The backbone.

Spinal cord The main nerve tract in vertebrates. It is normally enclosed within the spinal column.

Spindle tree A small tree of Eurasia and North Africa. It never grows more than 6 metres tall. It has lance-shaped leaves and tiny greenish flowers followed by beautiful pink and orange fruits. Its hard, smooth wood was used to make spindles, skewers and knitting needles.

Spinetail A group of swifts with short spiky tails. One of the commonest is the chimney swift of North America. Another species, the brown-throated spine-tail, is probably the fastest flying bird, reaching speeds in excess of 160 km/h.

Spiny anteater Also called the echidna, this creature is one of the egg-laying mammals or monotremes. It looks rather like a large hedgehog. Spiny anteaters are found in Australia and New Guinea. They are burrowing creatures which feed almost entirely on termites.

Spiracle An opening in the body wall of many arthropods (e.g. insects). Spiracles lead to the tracheae (breathing tubes) which admit air to the tissues. The name is also applied to a pore in front of the gill slits of some fishes.

Spiraea Any of 80 species of slender deciduous shrubs of the north temperate zone. They have small leaves and dense clusters of pink, red or white flowers. Dropwort and meadowsweet are closely related to spiraea.

Spleen A mass of spongy tissue found in the intestinal region of most vertebrates. The spleen and the lymphocytes that it produces play an important part in defending the body against infection. The spleen also acts as a store for red blood cells, releasing them when necessary.

Spleenwort Any of about 300 species of ferns which grow in rock crevices.

Sponge Multi-celled animals of the order Porifera which live in shallow seas, particularly in warmer waters, attached to the bottom. Some live singly, others are joined together to form colonies. Often the body wall contains a skeleton made up of branched sandy or chalky structures called spicules. The sponge lives on the food and oxygen contained in the sea water which it takes in through pores in the body wall and expels through one or more larger openings.

Spoonbill Long-legged wading birds related to ibises. The beak is flattened and broadens out at the tip to form a 'spoon'. Spoonbills are up to 75 cm long and are generally white. They are found in all continents, living in swamps and marshes where they feed on small aquatic creatures.

Sporangium, *see* SPORE

Spore A true spore is a tiny non-sexual reproductive body found in all groups of plants but especially in lower plants such as fungi, mosses and ferns. These all reproduce by scattering dust-like spores. Spores develop in a sac called a sporangium (plural sporangia). Sporangia are clearly visible in ferns as little brown patches under the fronds.

Spotted deer, *see* AXIS DEER

Spotted salamander, *see* FIRE SALAMANDER

Springbok A South African gazelle famous for its 'pronking' — the habit of leaping into the air. Pronking is a feature of all gazelles, but the springbok repeatedly leaps 3 metres or more into the air when alarmed or simply in play.

Springtail Primitive wingless insects, rarely more than half a centimetre long, which are able to leap some centimetres into the air using a forked 'spring' at the hind end of the abdomen. Springtails are found all over the world. They are particularly common in woodland leaf litter.

Spruce Any of about 40 coniferous evergreen trees of the north temperate zone. Spruces grow in a tall, pyramid shape. Their cones hang down, unlike those of FIRS, and they have short, four-sided needles which grow in short spurs on pegs on the twigs. They produce strong, straight-grained timber.

Spur The slender, hollow projection of a petal, often containing nectar.

Spurge The name of a family of over 7000 species of herbs, trees and shrubs. Most produce a milky juice, and they include CASTOR OIL and RUBBER TREES. Few have conspicuous flowers, but some, such as the poinsettia, have brilliant bracts. Most are poisonous, but the family also includes CASSAVA, an important tropical food plant.

Squash Any of several trailing vines producing large, fleshy fruits. They are in the GOURD family and include MARROW and PUMPKIN. Squashes are native to the Americas. They have large rough leaves and yellow flowers.

Squat lobster A small group of animals resembling lobsters but more closely

Springbok

related to hermit crabs. The body is stouter than a true lobster and the abdomen is tucked underneath. Squat lobsters never measure more than 15 cm in the body. They are confined to coastal waters in Europe.

Squid A group of marine molluscs related to the octopuses and cuttlefish. They have a large head bearing two prominent eyes, eight arms and two longer tentacles. The body tapers towards the rear and contains the vestigial mollusc shell. Squids range in size from a few centimetres to around 18 metres (including the tentacles). They are found in all oceans, the longest living at depths of 200–400 metres.

Starch, *see* CARBOHYDRATE

Starfish A typical starfish consists of five flattened arms radiating from a small central body on the underside of which is a toothless mouth. The arms, which can number as many as 50, are always symmetrically arranged. Starfishes with an irregular shape are those which have lost one or more arms and are regrowing or regenerating them. Starfishes are found in all seas, most living in shallow waters. They are carnivorous creatures: some prey on bivalve molluscs such as oysters and mussels, gradually pulling the two halves of the shell apart with their tube feet and then digesting the animal.

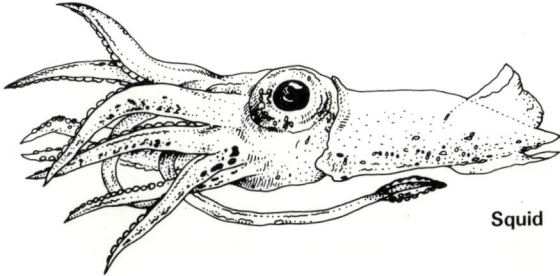

Squid

Squill The English name for the various species of SCILLA.

Squirrel In its widest sense this name is applied to any rodent of the family Sciuridae. But it is often restricted to the familiar tree squirrels — the grey and red squirrels and their relatives. They live in most parts of the northern hemisphere. They differ from the ground squirrels in having bushy tails about as long as the head and body. They are remarkably agile in the trees, although they often wander over the ground as well.

Squirrel monkey A small South American monkey with an almost human face. Squirrel monkeys live in large troops led by the females. The males tend to lead a solitary existence outside the mating season. The squirrel monkey's habitat ranges from the tree-tops to the forest floor and their food varies from fruit to insects.

Stamen The male part of a flower which carries the pollen-producing anther.

Stargazer Named for the fact that its eyes are set on the top of its head, the stargazer is a marine fish which buries itself in the sand of the seabed until just the eyes and the nostrils are showing. Capable of generating an electric shock of 50 volts, and armed with poison spines as an additional defence, the stargazer waits in relative security for

Squirrel

small marine creatures to swim by. Stargazers are found in many parts of the world. The largest is the northern stargazer of the north-western Atlantic which reaches 55 cm in length.

Starling A group of birds found in most parts of the Old World. One of the most familiar is the common starling which lives in flocks. During the day the flocks spread out to feed, but in the late afternoon they join together in noisy groups, sometimes more than one thousand strong, which whirl through the sky like clouds of smoke before roosting in favourite trees. Starlings are largely beneficial in agricultural areas since they tend to feed on insect grubs, but they have become serious pests in towns and cities. The common starling has been introduced into other continents. In North America it has spread the length and breadth of the continent during this century.

Steamer duck The name given to three large species of duck living in South America. Two of the species — the Magellan steamer duck and the Falkland Islands steamer duck — are flightless. The third, the flying steamer duck, is the smallest. These ducks are named for the way that they rush across the surface of the water, propelled by their wings and legs, throwing up sheets of spray. Steamer ducks are found in coastal waters.

Steinbok A dainty antelope with large eyes and ears which lives in scrub and bush country of southern and eastern Africa. It is rarely more than 50 cm high at the shoulder and has a coarse reddish coat. When alarmed, the steinbok's reaction is to lie hidden in the grass with its neck outstretched. Only if this fails will the animal run away.

Stem The stem of a plant performs two basic functions: it supports the leaves and flowers and it carries water and food from place to place within the

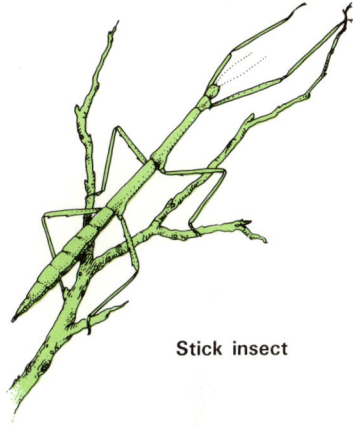

Stick insect

plant. The typical flowering plant stem is cylindrical and may be soft (herbaceous) or woody. It is usually branched and leafy. The point at which a leaf joins the stem is called the node.

Sterile (1) Unable to reproduce sexually. (2) Uncontaminated by living organisms.

Sternum The vertebrate breastbone.

Steroid Any of a number of important organic compounds that play a vital part in many bodily activities. Vitamins and hormones often contain steroid molecules.

Stick insect Long slender insects which closely resemble twigs, blades of grass or leaf ribs. There are many species, mostly found in South-East Asia, though a few species are native to southern Europe. The majority are wingless. Stick insects are often kept as pets.

Stickleback A group of small marine and freshwater fishes living in temperate waters of the northern hemisphere. The three-spined stickleback or tiddler is the most widespread. In the breeding season the male stickleback builds a nest from plant materials. The female lays her eggs in the nest and the male guards it until the young fish are ready to swim away.

Stifftail A group of ducks named for the stiff feathers that make up their tails. They have a short thick neck, a broad bill and short wings. They are truly aquatic birds, with the legs set

Starlings live in vast flocks, often of thousands of birds, and in flight they have the remarkable ability to suddenly change direction in unison.

Stigma

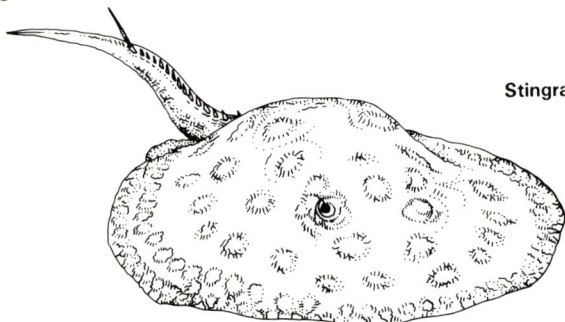

Stingray

well back under the body, and can only shuffle around awkwardly on land. The ruddy duck of America and the white-headed duck of southern Europe and central Asia are two well-known species.

Stigma The surface of a flower's carpel that receives pollen. It may be sticky or feathery.

Stilt A long-legged black and white wading bird related to the avocet. Stilts probe the mud with their long, slender beaks and feed on a wide variety of aquatic animals. They are found close to water in many parts of the world.

Stingray A group of flattened fishes related to skates which have a whip-like tail bearing a saw-edged poison spine. Stingrays live on the seabed in shallow tropical and temperate waters. They inflict an extremely painful wound on people unfortunate enough to tread on them.

Stinkbug, see SHIELDBUG

Stinkhorn A FUNGUS whose fruiting body is a tall white column topped by a slimy conical cap. It smells like bad drains and this attracts flies which carry away the spores.

Stipe A stalk — especially of toadstools and of algae.

Stipule An outgrowth of the base of a leaf-stalk. Often leaf-like but sometimes modified as spines.

Stoat A small relative of the polecat found over much of the northern hemisphere. It has a slender lithe body and relatively short legs. Stoats move in a series of bounds, their bodies undulating in a snake-like manner. The coat is chestnut brown above and white beneath, but in northern regions it

turns completely white in winter except for the black tip of the tail. The fur is then known as ermine. The stoat is a carnivorous animal feeding on rabbits, rodents and birds.

Stock The name of several strongly-scented wild and garden plants, mostly annuals with white or pink flowers, and of Eurasian or African origin. They include Virginian stock, from the Mediterranean, with small cross-shaped flowers, and night-scented stock whose slightly bigger flowers open at night. Stocks are members of the cabbage family.

Stolon A horizontal stem that can take root at the nodes and so give rise to new plants.

Stoma One of the pores in a leaf surface (plural stomata).

Stomach The expanded region of the alimentary canal where food is mixed and some digestion takes place.

Stonecrop The popular name of the sedum, a succulent herb of which there are about 120 species growing in rocky and sandy places. The common or bit-

Stoat

ing stonecrop, growing on walls or near the sea, is about 10 cm tall and has yellow, star-like flowers.

Stone curlew Also known as thick-knees, the stone curlews are a small group of waders found in many parts of the world with the exception of North America. Though technically waders they are normally found well away from water, particularly in sandy areas. They are strong runners and if danger threatens they prefer to run away rather than take to the wing. Stone curlews are nocturnal birds that feed mainly on insects and other small invertebrates.

Stonefish One of the ugliest and most poisonous of fishes, with a scaleless body covered in warts and 13 poisonous spines on the back. Stonefishes are found in tropical waters of the Indian and Pacific oceans, especially near coral reefs.

Stonefly A small group of insects that are placed in an order of their own. They have delicate, transparent wings and two long thread-like feelers protruding from the rear end. The larvae are aquatic and are often found under stones in fast-flowing streams.

Stork A family of large black or white birds that inhabit marshes and plains in most parts of the world. Adult storks have no voice but they can clap their beaks and make a loud noise. They hold their long beaks downwards when resting but extended horizontally when in flight. Storks feed chiefly on snakes, lizards, frogs, fishes and worms.

Storksbill A herb with beaked fruits, which give it its name. Storksbill is a native of Eurasia and North Africa and a member of the geranium family. It has much-divided oblong leaves and pink or white flowers. There are several species.

Storm petrel A small swallow-like bird of the open seas that can often be seen flitting to and fro across the wake of a ship, feeding on small animals thrown to the surface by the ship's propeller. The storm petrel is so named because sailors once believed that its arrival heralded a storm. There are a number of different species: one of the best known is Wilson's storm petrel which

breeds in the Antarctic region and migrates as far north as Newfoundland.

Strawberry A low-growing herb producing large, succulent red fruit. The fruit is a false fruit formed from the swollen receptacle. The true fruits are the woody pips on the surface. The leaves have three lobes and the flowers are white or pink. It spreads by means of runners.

Strawberry tree A small evergreen tree of western Europe. It has toothed, lance-shaped leaves, small flask-shaped flowers, and small rough fruits the colour of strawberries.

Stork

Stridulation The production of sound in certain animals by rubbing one part of the body (the scraper) against another (the file). It is particularly well developed among grasshoppers and crickets. As a rule, only the male stridulates. Short-horned grasshoppers have a row of pegs on their hindlegs and they rub these against a vein on the wing. Other species rub parts of the wings together.

Sturgeon A group of long-snouted fishes having features in common with both sharks and bony fishes. Most species live in the sea and return to fresh water to breed. The largest is the Russian sturgeon or beluga which may exceed 4 metres in length. Caviare is the egg-mass of the female.

Succession The progressive changes in the vegetation of a habitat from first colonization to the attainment of the CLIMAX VEGETATION associated with a given environment. A hillside after a landslip will first be colonized by lichens and mosses, then by grasses and other herbs, finally by shrubs and trees. Ponds gradually become filled as reeds and other water plants die and accumulate on the bottom, allowing terrestrial plants to encroach.

Succulent Fleshy: of fruits, or of plants which store water (XEROPHYTES).

Sucker A shoot arising from a root or underground stem, usually some way away from the main stem. It develops its own roots and becomes a separate plant.

Sucrose Cane sugar — a food reserve found in many plants.

Sugar beet A biennial kind of BEET rich in sugar. It has long, bright green leaves and tiny pinkish flowers, although the plants are normally harvested before they flower. The crown or upper part of the root is white and yields the sugar. The leaves may be eaten and are used for animal feed when the roots are harvested.

Sugar cane A giant perennial GRASS, originating in New Guinea and now grown in all tropical regions. The hollow stems, up to 5 metres tall and 5 cm in diameter contain a rich sugary juice. It is one of the main sources of sugar and syrup.

Sumach Any of about 120 species of shrubs from the warmer regions of Eurasia and North America. Some are slightly poisonous, others can be used medicinally. Sumachs have resinous sap, large compound leaves, small flowers and clusters of small berries. Poison ivy and the lacquer tree are species of sumachs. The name is also given to a material used in tanning and dyeing that consists of dried powdered leaves and flowers of various species of sumachs.

Sun bear Also called the Malay bear, this animal is the smallest of the bears. It is a stocky creature, about 1 metre long and 60 cm high. It has short black hair, with a grey muzzle and a distinct white or yellowish crescent across the chest. The sun bear lives in the forests of South-East Asia and spends most of its time in the tree-tops.

Sunbird The Old World equivalent of the hummingbird, sunbirds are found in the tropical forests of Africa and Asia. The two groups both have similar brilliant plumage and feeding habits, but they are not related.

Sundew Any of 90 species of small carnivorous plants that live in marshy and boggy places. They have flat leaves covered with hairs bearing drops of sticky fluid. Insects are attracted to the glistening fluid and are trapped on it. The leaves then close around the insects and digest them.

Sunflower Any of 60 species of tall North American composite plants bearing large, sun-like flower-heads. They have big coarse leaves and some grow up to 5 metres tall, although many are only annual plants. The seeds of the common sunflower are an important source of oil. The Jerusalem artichoke is a perennial kind of sunflower.

Suni One of the smallest of the antelopes, just 40 cm high, the suni is found in the lowlands of eastern Africa. A close relative, the royal antelope of western Africa, is even smaller, only 25 cm high at the most.

Sun bear

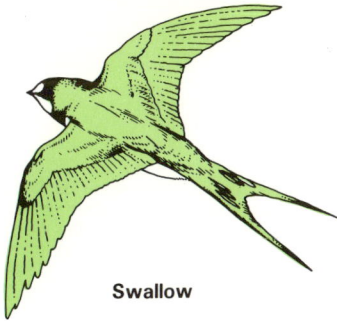

Swallow

Surgeonfish Colourful fishes that get their name from the razor-sharp bony keels at the base of the tail. Surgeonfishes are found around coral reefs of the Indian and Pacific oceans. They grow to lengths of about 60 cm and have deep bodies flattened from side to side.

Surinam toad A small South American toad with an unusual life history. The eggs stick on to the spongy skin of the female's back and are carried around with her. A pouch and lid develops around each egg and the whole larval development takes place in the pouch. After about 4 months tiny toads crawl out of the pouches and swim away.

Suslik, *see* GROUND SQUIRREL

Swallow The bird known in Europe as the swallow and in North America as the barn swallow is the most widely-spread member of the family Hirundinidae which includes swallows and martins. It is a migratory bird which breeds in Europe, Asia and North America and winters as far south as southern Africa and Australia. Swallows generally return to the same spot year after year to breed, navigating thousands of kilometres from their winter quarters with uncanny accuracy. Swallows are insect-eating birds that build cup-shaped nests from mud and grass in barns and outhouses.

Swallow plover, *see* PRATINCOLE

Swallowtail A group of butterflies named for the long 'tails' on their hindwings which resemble the forked tails of the swallow. Swallowtails are found in most parts of the world,

though the majority of the species live in the tropics. They include some of the most beautiful and largest butterflies, and are closely related to the tropical birdwings.

Swamp deer The largest South American deer, standing one metre high at the shoulder. As their name suggests, these deer live mainly in marshy areas and their widely-spreading hoofs are an adaptation to walking on soft ground. Swamp deer are found in south-western Brazil.

Swamp hen, *see* GALLINULE

Swan A group of long-necked water birds closely related to geese. Most familiar is the mute swan, which is all white with an orange bill. It is a native of parts of Eurasia but has been taken to various other parts of the world. Whooper swans and Bewick's swans occur in Europe and North America. They have black or black and yellow bills. Black swans live in Australia, and black-necked swans live in South America.

Black swan

Mute swan

Sweat gland An organ of the skin of mammals which is important to temperature control and water-loss.

Swede Also known as Swedish turnip, this is a root vegetable of the cabbage family. The sweet flesh is yellow and very nutritious for both people and farm animals. It can be distinguished from the turnip by its colour and its long neck which bears leaf scars. In America the swede is called rutabaga.

Swift

Sweet cicely, *see* CHERVIL

Sweet pea An annual climbing herb related to the ordinary PEA. It is grown for its butterfly-like fragrant flowers, which are white, pink or purple.

Sweet potato A trailing perennial herb related to MORNING-GLORY. It has large leaves of varying shape and funnel-shaped flowers. It produces a large, tuberous root weighing between 1 and 5.5 kg.

Swift Superbly adapted to an aerial life, swifts rarely land on the ground. Their long narrow wings are suited to high-speed flight and they include the fastest of all birds. Swifts are insect-eating birds and those that live in temperate regions migrate to tropical regions in the winter. There are two families: the crested swifts and the true swifts, the latter including the SPINETAILS.

Swift lizard, *see* FENCE LIZARD

Swim-bladder A gas-filled sac lying above the gut of bony fishes. The walls secrete or absorb gas to maintain the buoyancy of the fish as it dives or rises through the water.

Swordfish A striking blue and silver fish in which the upper jaw is carried forward to form a sword-like beak. The dorsal (back) fin is very tall. The fish is normally between 2 and 3 metres long, including the sword which is up to about 1 metre long. Swordfishes live mainly in the warmer seas and feed on smaller fishes.

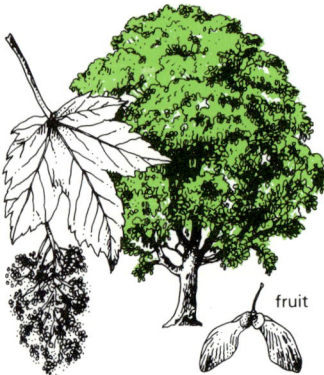

Sea anemones frequently live on hermit crabs, with an advantage to both — a symbiotic relationship.

Swordtail A small brightly coloured aquarium fish related to the carp. The sword is an extension of the lower part of the tail fin in the male. Swordtails live in fresh water in Central America.

Sycamore A species of MAPLE native to Eurasia. It can grow up to 35 metres tall. It has large leaves divided into five lobes, and long clusters of tiny greenish flowers. In America a species of PLANE is called sycamore.

Symbiosis The association of two different organisms in which they both gain advantages: e.g. birds feeding off the parasites that infest a crocodile's skin and mouth have a symbiotic relationship with the crocodile.

Syrian bear, *see* BROWN BEAR

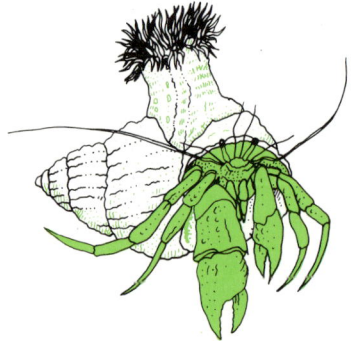

fruit

Sycamore

T

Tactile Concerning the sense of touch.

Tailor-bird A group of warblers living in southern Asia and Australia. They are small birds with inconspicuous plumage, and they feed entirely on insects. Their nests are made by stitching leaves together to form a pouch which is then filled with soft fibres or grasses.

Takahe A rare flightless bird of New Zealand. It is related to the coot and stands about 45 cm high. The wings and body are green and blue and the beak is pink and scarlet. The bird lives in swampy regions and feeds entirely on the leaves and seeds of a plant called snow grass.

Takin A heavily built mammal, half goat and half antelope, standing about one metre high. The male is usually yellowish, with a dark stripe along the back. Females are greyish and the young are black. Takins live in the mountains of western China and feed mainly on bamboo.

Tallow tree A Chinese tree whose seeds are coated with a tallow-like substance, used for making candles and soap. The tree has pointed oval leaves which yield a black dye, and spikes of greenish flowers. The wood is hard.

Tamandua, *see* ANTEATER

Tamara, *see* LOTUS

Tamarind An evergreen tree of the Old World tropics. It grows up to 23 metres tall and has leaves divided into as many as 40 leaflets. Its tiny yellow flowers produce long pods containing an acid pulp which is used in chutneys and also as a laxative.

Tamarisk The name given to over 60 species of trees and small shrubs native to Europe and India. They grow in salty deserts and on seashores, and have scale-like leaves on long slender branches, and clusters of small white or pink flowers. MANNA appears on one species.

Tanager A large group of medium sized birds related to the buntings. Most are brilliantly coloured and nearly all live in the forests of tropical America. They feed on fruit and insects.

Tansy A perennial, strong-scented Eurasian herb. It has angular stems about 90 cm tall, feathery leaves, and clusters of dull yellow composite flowers. It used to be used as a medicine and to flavour food.

Tapetum A reflecting layer behind the retina of many nocturnal animals. It reflects back onto the retina light that would otherwise be lost. The tapetum is responsible for the coloured reflection from the eyes of cats and other animals when a torch is flashed at them at night.

Tapeworm A group of parasites which live as adults in the intestines of vertebrates, including human beings. The head end is armed with hooks or suckers to anchor the animal to the intestine wall where it shares the food of its host.

Tapioca, *see* CASSAVA

Tapir A stoutly built mammal related to both the rhinoceros and the horse. It is about 1.5 metres long and stands up to 1 metre at the shoulder. The snout is prolonged into a short trunk, and the body is covered with short bristly hair. Tapirs live in the wet tropical forests of South-East Asia and South and Central America.

Tap root, *see* ROOT

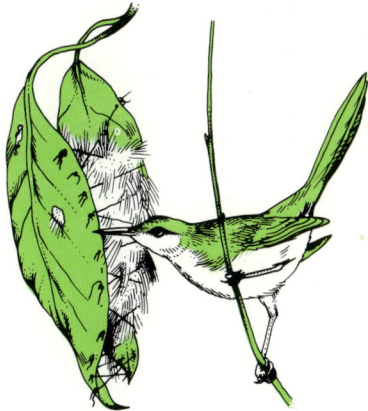

Tailor-bird and nest

Tarantula This name is popularly applied to almost any large spider, especially to the hairy bird-eating spiders of South America. The true tarantula, however, is a wolf spider found in southern Europe. It is a greyish spider about 2.5 cm long and with a mildly painful bite.

Tardigrade Also known as water bears, tardigrades are tiny animals, rarely more than 12 mm long, that are classified with the arthropods. They feed on plant juices and live mainly among damp moss and lichens.

Tare Any of several short annual herbs. It has weak stems, leaves divided into leaflets, and blue or white flowers. It grows among long grass. It is related to vetch, and like it has seeds in pods.

Tarpon This is a marine and fresh-water fish with many common names, including big-scale, silverfish and silver king. It is up to 2.5 metres long and lives along the tropical and sub-tropical coasts of the Atlantic. The tarpon reacts violently to being hooked and it is a big-game fish for the sea angler.

Tarragon A perennial herb whose fragrant leaves are used for flavouring. It is a native of southern Europe. It grows about 60 cm tall, and has narrow

Tarsier

oblong leaves and green flower heads. It is related to wormwood and is a member of Compositae family.

Tarsier The three species of tarsier are found only on certain islands in the East Indies. They are small nocturnal primates, the head and body of the largest being only about 15 cm long, excluding the tail which is over 20 cm in length. Tarsiers have thick woolly fur, enormous eyes and large rounded sucker-like discs on their fingers and toes. They spend much time clinging to vertical tree-trunks and feed mainly on insects.

Tarsus The ankle region of the hind-leg.

Tasmanian devil A carnivorous marsupial now thought to be confined to Tasmania. It is stockily built, rather bear-like, and up to 1 metre long including the 60-cm tail. Its coat is black with white patches on the throat, shoulder and rump. It has large pink ears and its powerful teeth give it a rather fierce expression.

Tasmanian wolf, *see* THYLACINE

Taste One of the chemical senses. Unlike the olfactory sense (the sense of smell), which detects materials at a distance, the organs of taste must come into contact with the material in a concentrated form. When various materials affect the taste organs the appropriate messages are sent to the brain. We do not know how the taste organs distinguish different substances and send the right messages to the brain, but it is likely that each taste organ is sensitive to just one kind of material. This certainly seems true of the human tongue, which can detect only four basic tastes. Sweet and salty things are detected mainly by the tip of the tongue; acidic (sour) things are detected by the sides; and bitter things affect the back of the tongue. The mouth-watering tastes of favourite foods are not really tastes at all — they are scents or flavours detected with sense organs in the nose.

The taste organs of vertebrate animals are nearly all found on the tongue. They consist of little clusters of nerve endings wrapped around patches of spongy tissue known as taste buds. Each sends a nerve fibre to the brain. Taste buds also occur on the lips of fishes. Materials must dissolve in the surrounding fluids before they can affect a taste bud. Among the invertebrate animals the organs of taste can occur in the most unlikely places. Many insects, for example, have taste organs on their feet, as well as on the sensory palps around the mouth. They can tell, just by landing on a plant, whether it is good to eat or suitable for laying their eggs on.

Tawny owl A common owl of Europe and parts of Asia. It is about 45 cm in length with a large head and a greyish facial disc. The tawny owl has a mournful call which includes the 'tu whit' and 'to whoo' sounds. Its large eyes are adapted to night hunting but it also has a keen sense of hearing. The diet of the tawny owl consists mainly of small mammals such as shrews and voles.

Tawny owl

Taxis The movement of a whole organism in response to some stimulus, such as the smell of food which causes an animal to move towards the source of the smell.

Taxonomy The study of the naming and classification of animals.

Tea An evergreen shrub or small tree native to southern and eastern Asia. It can reach a height of 9 metres, but is cultivated as a bush. It has long elliptical leaves and white flowers, and pro-

flowering head

seed head

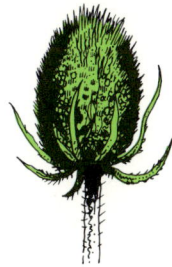

Teasel

duces woody fruit. The tea that is drunk is produced from young leaves and buds.

Teak A large deciduous tree of southern Asia, related to verbena. It grows up to 45 metres tall, and has leaves up to 60 cm long. It produces white flowers and woolly fruits. The timber is strong and oily and exceptionally durable.

Teal A group of small ducks found in many parts of the world. The common teal of Europe and North America has a chestnut and green head in the male, and both sexes have green and black markings on the wings.

Teasel A tall prickly herb native to Eurasia and the Canaries. It grows up to 1.2 metres high, and has long saw-edged leaves and thistle-like flowerheads. The dried flower-heads have hooked bracts which have long been used to 'tease' out wool and to raise the nap on cloth.

Telson The last segment of the abdomen of arthropods. It does not always develop — insects do not have one for example — but the scorpion sting and the fan-shaped tail of lobsters are examples of a well-developed telson.

Tendon A bundle of fibres connecting muscles to bones.

Tenrec Small insectivorous mammals, up to 35 cm long, related to shrews and hedgehogs. They live only in Malagasy (Madagascar). There are a number of species, some looking like hedgehogs, some like shrews, and they behave very much like the animals they resemble.

Terebinth A small tree of the Mediterranean region, related to the pistachio. It grows up to 9 metres tall, and has leaves broken up into lance-shaped leaflets, and greenish flowers. Incisions in the bark yield a resinous fluid from which turpentine is made. It is also called the turpentine tree.

Tergum The thickened dorsal region of the arthropod cuticle.

Termite Although sometimes called white ants, the nearest relatives of the termites are cockroaches. Termites are mainly tropical insects which live in huge colonies consisting of soldiers, workers, a king and a queen. Many species build rock-hard nests from earth and saliva which rise as much as 6 metres from the ground. Others live in wood, feeding on the timber: these cause tremendous damage to houses and furniture.

Tern A group of sea-birds related to the gulls. Their wings are narrow and tapered and their tails are forked. The bill is narrow and the legs are short. Most of the species are black and white and range from 25 cm to 40 cm in length. Most terns live in the tropics, but there are terns in all parts of the world.

Terrapin This name is loosely used for any tortoise or turtle living in fresh water. More strictly, however, it means the diamondback terrapin found in the coastal waters and estuaries of the United States and Mexico. This animal gets its name from the diamond-shaped pattern on its shell. It is up to 20 cm long.

Territory An area inhabited or dominated by an animal or a family of animals, especially for the purpose of reproduction. Territory is marked out by scent, or by singing in the case of birds, and intruders of the same species are attacked. Territory is a way of population control for it prevents too many animals from breeding in one area.

Testis The male organ (gonad) producing sperms and various sex hormones.

Tetra, *see* CHARACIN

Tetrapod A term applied to any four-footed animal.

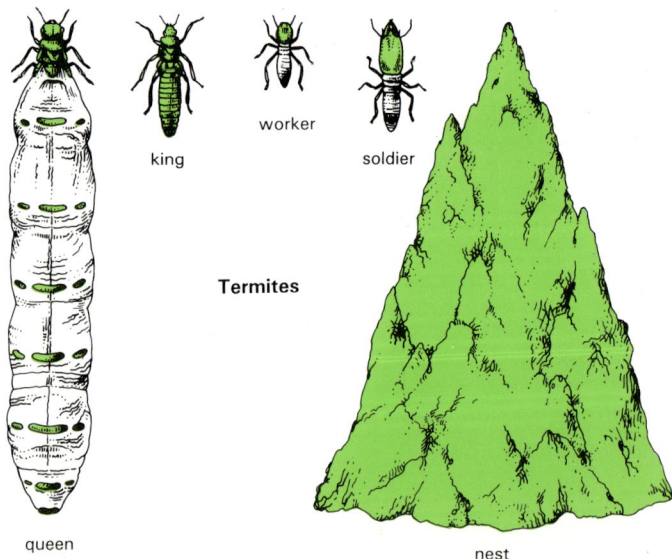

king

worker

soldier

Termites

queen

nest

Thresher sharks

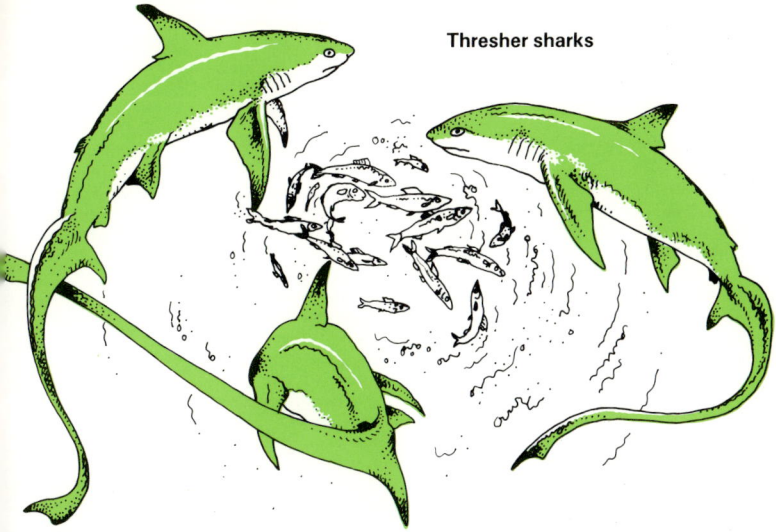

Thallophyte Any of a large division of the plant kingdom (Thallophyta) which contains the most primitive plants — algae, fungi and lichens, and also the bacteria and slime fungi. The body of thallophytes is simple in that it is not divided into root, stem and leaf. This simple body is called a thallus.

Thick-knee, *see* STONE CURLEW

Thistle The name of various spiny-herbs found in Eurasia and North America. They have spiny leaves and often very prickly stems. The composite flower-heads are roundish, red, purple or yellow. Each seed has a para-chute of silky hairs (thistledown).

Thistle butterfly, *see* PAINTED LADY

Thomson's gazelle, *see* GAZELLE

Thorax (1) The region of the ver-tebrate body surrounded by ribs (when present) and enclosing the heart and lungs. (2) The central region of the insect body consisting of three seg-ments and normally carrying three pairs of legs and two pairs of wings.

Thorn A sharp, woody structure found on certain plants, strictly one that is a modified stem as in the hawthorn. Other pointed structures, such as those of the acacia which are modified sti-pules, are called spines. Thorns are sometimes called stem-spines. The prickles of roses are outgrowths of the stem and are called emergences.

Thorn apple The popular name for the datura, any of a group of poisonous herbs. Daturas have toothed leaves that smell unpleasant, large funnel-shaped flowers, often brightly col-oured, and prickly fruit. One common American species is called jimson weed. It has white flowers and has spread to many parts of the world.

Thorny devil, *see* MOLOCH

Thresher Thresher sharks, also called fox sharks, are found mainly in the warmer seas of the world. They reach lengths of up to 6 metres but half of this is tail. The sharks feed on small fishes such as herrings and pilchards, which they 'round up' and then stun by fierce lashing of the long tail.

Thrips A tiny black insect with a nar-row body and usually with narrow fringed wings. Thrips are extremely common and can be found in almost any flower in summer. They fly about and often get in our eyes and hair. They hibernate for the winter and often come into houses where they hide behind picture frames, loose wallpaper and in other crevices. Thrips feed by sucking plant juices and some are real pests, especially of cereals.

Thrush

Thrush A name originally applied to two European birds — the song-thrush and the mistle-thrush. It is now also used to describe a large number of related birds; the American robin, the European robin, the blackbird and the nightingale to name but a few. Both the song-thrush and the mistle-thrush have brownish backs and speckled underparts, but the mistle-thrush is the larger of the two, with a whiter breast and greyer-toned plumage. The song-thrush is yellow on the upper breast with a yellowish patch on the flanks.

Thrush

Thylacine A dog-like flesh-eating marsupial, also called the Tasmanian wolf. It is about 1 metre long, with a 45-cm tail. Although the front half is dog-like, the hind quarters and tail are much more like those of a kangaroo. The fur is light brown, with darker stripes across the hind part of the back. Thylacines were once common in Australia, but now are extremely rare and survive only in the wilder parts of Tasmania.

Thyme A fragrant herb of the mint family. It is low growing, with narrow leaves and small purple flowers. The leaves are used for flavouring. There are several species.

Thyroid gland A gland found in the throat region of all vertebrates. In man it has two lobes surrounding the windpipe just below the larynx or voice-box. The gland produces several HORMONES, the main one of which is called thyroxine. This essentially controls the rate of respiration in the body, and therefore affects the whole metabolism of the animal.

Tibia The shin bone — the larger of the two bones in the lower hindleg.

Tick Small blood-sucking creatures related to spiders and, more closely, to mites. There is a small head and a larger, leathery body. The head carries mouthparts which pierce the host — usually a mammal — and withdraw blood. The tick stays put for long periods and gradually swells up as it sucks blood. More serious than the actual blood sucking is the fact that the ticks spread diseases.

Tiddler, *see* STICKLEBACK

Tiger The largest of the cats, the tiger is a magnificent striped beast ranging from Bali in Indonesia to Iran and Siberia. The tigers of Siberia and northern China are very large and may reach almost 4 metres in length, although 3 metres including tail is a good average for tigers from other areas. There are several races, differing in colour, pattern and size. Tigers live in a variety of habitats, from mountains to dense forests. They do not climb well, but they are good swimmers. They eat a variety of animals, from fish to antelopes.

Tiger moth A group of often gaudily coloured moths. Their larvae are always hairy and are often called woolly bears. The garden tiger moth has cream coloured forewings with dark markings and red hindwings with blue-black spots.

Tilapia This name is given to many freshwater fishes in Africa and the Middle East. They belong to the cichlid family and have large heads and deep bodies compressed from side

Thylacine

Tiger moth

to side. Most are between 20 cm and 30 cm long and they live in lakes and sluggish rivers. Many tilapias carry their eggs and young in their mouths.

Tissue A patch or group of similar cells in an organism which perform the same function, e.g. muscle cells, liver cells, etc.

Titmouse A group of small woodland birds, often simply called tits, found over much of the northern hemisphere. The blue tit and the similar but larger great tit are familiar garden birds, well known for their acrobatic skills at the bird table and the way they learn to peck through milk bottle tops to feed on the cream.

Toad The name given to a number of amphibians belonging to the same genus as frogs. Toads tend to be more terrestrial than frogs, with a dry warty skin, though there are many exceptions. The common toad of Europe reaches 15 cm in length, with the male being slightly smaller than the female. Other toads include the FIREBELLY, MIDWIFE TOAD, NATTERJACK TOAD, SPADEFOOT TOAD and SURINAM TOAD.

Toadflax Any of several Eurasian herbs with tubular flowers resembling those of a snapdragon. They are yellow and orange, and a trailing variety has bluish flowers. The leaves are small and narrow.

Toadstool The name normally used for any fungus with the familiar umbrella-shaped fruiting body, particularly inedible or poisonous species, but sometimes applied to other large fungi as well.

Tobacco Any of 66 species of herbs, mostly American but a few Australian.

Some are grown as ornamental garden plants, but one species, the common tobacco, is used for smoking. It stands about 1.8 metres tall. It has leaves up to 75 cm long and tubular pink flowers. The leaves contain the narcotic poison nicotine.

Tomato An annual herb related to the potato and nightshade, native to South America. It has large, hairy compound leaves, clusters of small yellow flowers, and produces large, soft succulent red or yellow berries which are treated as vegetables.

Tomb bat The name given to several kinds of insect-eating bat which roost in tombs in Egypt, though they are found in most other tropical parts of the world as well. The body is rarely longer than 10 cm and brown in colour, though several species known as ghost bats are completely white.

Tonsils Masses of lymphatic tissue located in the throat on either side of the soft palate. The adenoids are tonsils at the back of the nose. Tonsils may be a defence against harmful bacteria.

Tooth A bone-like body in the jaws of most vertebrates concerned mainly with the capture and breaking up of food. The teeth of reptiles and other lower vertebrates are all of a simple conical shape, but mammalian teeth are of several different types. At the front of the jaws there are a number of chisel-shaped incisor teeth. These are followed by a stabbing or eye tooth —

Blue tit

Great tit

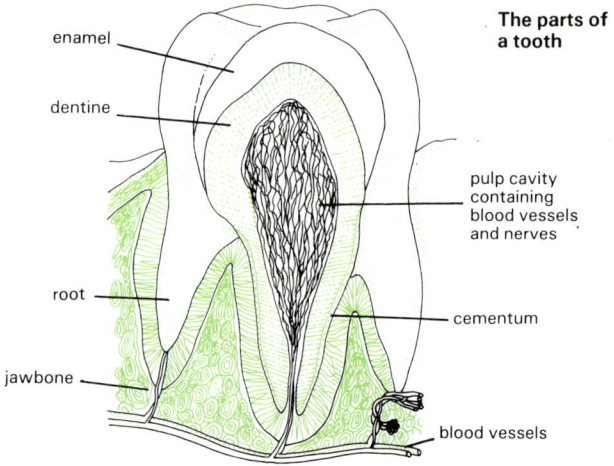

The parts of a tooth

enamel

dentine

pulp cavity containing blood vessels and nerves

root

cementum

jawbone

blood vessels

the canine — and then the grinding cheek teeth — premolars and molars. The number, structure and arrangement of teeth differ between animals that have different diets.

All teeth are constructed on the same basic plan. Projecting from the gum is the crown of the tooth. The part embedded in the gum and reaching into a socket in the jaw is known as the root. The body of the tooth is made up of a hard bone-like substance known as dentine. Inside this is the pulp cavity which contains blood vessels and nerves. Covering the crown of the tooth is a layer of enamel. Around the root of the tooth enamel is replaced by cementum, another bone-like material which fixes the tooth firmly in the socket of the jaw.

Tooth-carp A group of small freshwater carp-like fishes with comb-like teeth in the jaws and throat. Many of the tooth-carps are brightly coloured and are popular aquarium fishes. Guppies, for instance, are tooth-carps.

Topshell Cone-shaped sea snails found in most coastal waters but especially in the tropics. Many have decorative shells and the largest measure up to 15 cm across the base.

Torrent duck A South American duck unusual in living among the rapids and waterfalls of fast-flowing streams. Tor-

rent ducks are found in the foothills of the Andes throughout the length of South America.

Tortoise A heavily armoured reptile renowned for its slowness of movement and its long life-span.

Tortoises vary considerably in size: the familiar 'garden' tortoise of southern Europe and North Africa has a shell up to 30 cm long; the Galapagos giant tortoise can reach over 1.5 metres in length. When alarmed a tortoise can withdraw its head and limbs into the safety of its bony 'shell'. Records suggest that tortoises can live up to 150 years and possibly over 200 years.

Tortoise

Tortoiseshell butterfly The large and small tortoiseshell butterflies range across Eurasia, though the small tortoiseshell is far more common than the large tortoiseshell. Both are reddish-orange with black markings and the small tortoiseshell has blue crescents at the ends of the wings. The food plant of the small tortoiseshell larvae is the stinging nettle, that of the large tortoiseshell elm foliage. Tortoiseshells are related to the peacock and red admiral butterflies.

Toucan Large birds of tropical America with bright colours and huge bills, sometimes as long as the rest of the bird. Toucans live in the forests, usually going about in small flocks. They feed almost entirely on fruit, supplemented by some insects. The bills are probably also used for display.

Touch-me-not, *see* BALSAM

Trachea (1) The wind-pipe of tetrapods. (2) A breathing tube of air-breathing arthropods.

Tradescantia Any of 60 species of garden and house plants cultivated for their flowers or leaves. The flowers have three petals and are triangular in shape. They are usually white or purple. The leaves are oval, sometimes striped.

Tragopan A group of brightly coloured pheasants found in upland forests of southern and eastern Asia. The male bears two fleshy horns on the head and a bib-like wattle on the throat which are displayed during the breeding season.

Transpiration The loss of water through the leaves of a plant as water vapour.

Trapdoor spider Trapdoor spiders are related to the large, hairy bird-eating spiders, but instead of hunting for food they lie in wait for their prey. They construct silk-lined tunnels in the soil and fit them with lids of silk. The spiders then sit peering out of the doors waiting to pounce on passing insects. Trapdoor spiders live only in the warmer regions of the world. They are not dangerous, although some species can give a painful bite.

Traveller's tree A palm-like tree that grows in Malagasy (Madagascar). It has very large leaves, and the hollow

Toucan

stalks each hold about 0.5 litres of water. It has white flowers and edible seeds. Its name derives from the fact that it provided travellers with food, drink and shelter.

Tree creeper, *see* CREEPER

Tree duck Also known as whistling ducks, the tree ducks are large, goose-like birds related to swans. Despite their name these duck are less at home in trees than many other common ducks. Their nest is usually on the ground near water. Tree ducks are found throughout tropical and sub-tropical areas. There are eight species, all with whistle-like calls.

Trapdoor spider

Tree fern Any of several large species of ferns which resemble palm trees. They are most often found in tropical mountain regions. They can grow up to 12 metres tall and have fronds up to 9 metres long. They were common in prehistoric times, and their remains contributed to the formation of coal.

Tree frog The name given to a family of tree-dwelling frogs widely distributed with many different species. They have suckers on their toes and some have webbed hind toes. Some breed in water-filled tree holes but others come down to ponds to breed.

Treehopper, *see* HOPPER

Tree kangaroo Tree-dwelling marsupials related to wallabies found in Australia and New Guinea. They measure up to 75 cm long in the body with a furry tail about as long again. Though ungainly on the ground these animals scamper nimbly through the trees.

Tree of Heaven A fast-growing Chinese tree, also called ailanthus. It grows quickly to a height of 25 metres, and has long compound leaves, small greenish flowers, and seeds in twisted wings. It withstands pollution and is widely used as a shade tree.

Tree kangaroo

Tree porcupine The tree porcupines of North and South America differ in a number of respects from the Old World porcupines, but they are similar in possessing quills, though somewhat shorter, which can be erected against an attacker. The tree porcupines of South America have a long prehensile tail.

Tree shrew Small squirrel-like mammals living in South-East Asia. They are 20–40 cm long, about half of which is tail, and generally brownish. Tree shrews are generally included with the primates and it is thought that some creature of this type gave rise to the ancestors of monkeys, apes and man.

Trefoil The name applied to various clovers and related plants with compound leaves usually composed of three leaflets. They belong to the pea family. Bird's-foot-trefoil has five leaflets, although only three are obvious, and masses of red-tinged yellow flowers which give it its popular name of bacon-and-eggs.

Trichome A plant hair. Trichomes are found on most stems and leaves. Plants growing in exposed conditions frequently have a good covering of hairs which protect them from the cold or from the drying effects of sun and wind.

Triggerfish A group of small fishes with deep, almost triangular-shaped bodies, of which the head makes up at least one-third. They have an extraordinary range of markings in a variety of bright colours. Triggerfishes are found in tropical coastal waters, particularly near coral reefs.

Trogon A group of brightly coloured birds found throughout the tropics. Most brilliant of all is the QUETZAL of Central America.

Tropic bird Graceful sea-birds with two very long tail feathers. They look rather like terns, but are actually related to cormorants and pelicans. They are white with black markings and are up to 60 cm long without the

Tree frogs are able to change colour rapidly to avoid detection by their predators. The adhesive climbing-discs on the end of their toes enable them to cling to leaves.

Tuatara

tail plumes. The three species all live in the tropical regions and come to land only in the breeding season.

Tropism A bending movement of plants in response to a stimulus such as light or gravity.

Trout The classic game fish of Europe and North America, trout are closely related to salmon. Most species prefer cold-water streams or deep lakes. The European trout exists in several forms, including the brown trout, which stays in fresh water, and the sea trout, or salmon trout, which spends part of its adult life in salt water. Large numbers of trout are raised in commercial fish hatcheries.

Truffle Any of several kinds of edible fungus that produce their fruiting bodies underground. They are potato shaped, and grow near the roots of trees. Dogs and pigs are used to find them by scent. They are highly prized for their flavour.

Trumpeter Named for its deep reverberating call, this bird resembles a short-legged crane. There are three species living in the depths of the Amazonian rain forests. They are weak fliers and spend most of their time on the ground.

Trunkfish Also known as boxfishes, the trunkfishes are almost completely encased in bony boxes. Only the tail and the fins are unenclosed. The head is more or less conical and has a small mouth armed with strong crushing teeth. The eyes are large and placed near the top of the head. Trunkfishes are bottom-living fishes found in the warmer seas of the world.

Trypanosome A group of protozoans some of which are parasites and cause serious diseases such as sleeping sickness in man and nagana in cattle. The organisms are transmitted by TSETSE-FLIES.

Tsetse-fly African insects which carry the dreaded disease called sleeping sickness. The flies also carry a disease of cattle and horses called nagana. Tsetse-flies look rather like house-flies, but they suck blood. Unlike mosquitoes and horse-flies, both male and female suck blood.

Tuatara A lizard-like creature which is the sole survivor of a group of reptiles that became extinct 70 million years ago. Tuataras are confined to a few rocky islands off the coast of New Zealand, where they are strictly protected.

Tuber A food store formed from a swollen root or underground stem. The potato is a stem tuber but dahlia tubers are modified roots. Tubers are able to grow into new plants when detached from the parent plant.

Tubeworm A group of bristleworms that live in tubes of various kinds. They are found buried in the sand or attached to rocks and seaweeds on the seashore. The chalky white tubes sometimes found on the surface of stones are made by tubeworms.

Tulip tree

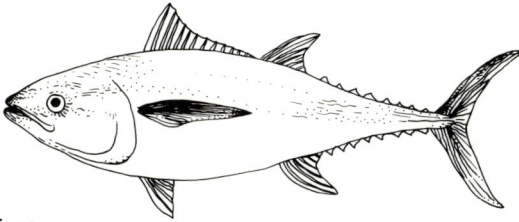

Tuna

Tubifex A mud-dwelling aquatic worm resembling a small red earthworm. There are a number of species, rarely more than 5 cm long, some living in fresh water, others in the sea. These worms are commonly used by aquarists as fish food.

Tufted duck, *see* POCHARD

Tulip A bulbous herb of the lily family, native to Eurasia. There are many species, all of which have long leathery leaves and a single, deep cup-shaped flower on a long stalk. The flowers occur in every hue except blue.

Tulip tree A tall deciduous tree of the magnolia family, native to North America. It may grow up to 60 metres high. It has four-lobed leaves with straight, notched ends, and showy tulip-like yellow flowers. The timber is easy to work.

Tuna Large oceanic fishes of which there are several species. The bluefin tuna or tunny of the Atlantic and Mediterranean is said to reach lengths of 4 metres, although most specimens are not much more than half of this size. Tuna are very streamlined and fast-swimming fishes with deep bodies, dark blue above and silvery beneath.

Tunny, *see* TUNA

Turaco Fruit-eating birds related to the cuckoos. Most are about 45 cm long, with short rounded wings and longish tails. They have strongly curved bills and all but one species have large crests on the head. Turacos live only in the African forests. Some species are called plantain-eaters.

Turbot A highly prized food fish with a broad, lozenge-shaped body up to one metre in length. The turbot is a flatfish found in the Mediterranean and the seas off western Europe as far north as the Arctic circle.

Turkey The wild turkey, a native of America, was domesticated by the Aztecs before the arrival of the Europeans. By the end of the 16th century it was a popular table bird at European banquets and was later reintroduced to the New World. There are two forms of wild turkey: one distributed across the United States and simply called the turkey, the other, known as the ocellated turkey, found in the forests of Central America.

Turkey vulture This bird is one of the 'New World vultures', with a wingspan of up to 2 metres. It is found throughout North and South America, where it is commonly known as the buzzard (although it is not, of course, closely related to the European bird of that name).

Turmeric A plant of southern Asia, also called curcuma. It is related to ginger. It has fleshy roots which when dried and powdered yield the condiment turmeric, an ingredient of curry powder also used as a yellow dye.

Turnip A biennial herb related to the cabbage. It has yellow flowers, and long bristly leaves which are sometimes eaten. The upper part of the root forms a sphere which is eaten as a vegetable. It is white or yellow, often with a green or purple tinge at the top.

Turnstone

Turnstone Small wading birds that get their name from their habit of turning over stones when searching for food. They are about 22 cm long, with short legs and a short, slightly up-turned bill. The common turnstone breeds all around the Arctic regions. The black turnstone breeds only in Alaska. Both species migrate far south for the winter.

Turnstone

Turpentine tree, *see* TEREBINTH

Turtle The turtles and tortoises make up the reptile order Chelonia. The name turtle is normally used for the marine and freshwater species and the name tortoise for the land-living species. These animals are easily recognized by the box-like shells in which they live. The shells are made of bone covered by horny plates. There are gaps for the limbs and the neck, and the head can usually be withdrawn into the shell by bending the neck. The land-dwelling tortoises feed mainly on plants, but the water-dwelling species are mainly carnivorous. All lay hard-shelled eggs.

Turtle-dove A small pigeon which breeds in most parts of Europe and winters in the southern hemisphere. It is about 25 cm long and mostly grey in colour with reddish-brown upper parts and blackish tail feathers.

Twaite shad, *see* SHAD

Twinflower A downy evergreen undershrub of the northern coniferous forests. It has long woody stems that creep along the ground, forming a mat, and slender stalks each bearing a pair of pink, bell-shaped flowers.

Twitch-grass, *see* COUCH-GRASS

Tympanum, *see* EAR

UV

Ugli A large West Indian citrus fruit with a wrinkled skin. It is a cross between an orange and a grapefruit.

Ulna One of the bones of the forearm — on the side of the 5th digit.

Umbel An inflorescence in which all the flower stalks come from the same point.

Ungulate A hoofed mammal. Ungulates are normally herbivorous and tend to live in herds. The hard hooves are well-suited to life on open grasslands.

Upas tree A large tree of south-eastern Asia, especially Java, and Africa. It has oval leaves and produces small berries. It has a milky latex that is highly poisonous, and was used by tribesmen to tip their arrows.

Urania moth A group of brilliantly coloured day-flying moths mostly confined to the tropics. In most species the underside is equally colourful. The largest and most brilliantly coloured species lives in Madagascar. The urania moths resemble swallowtail butterflies including the 'tails' on the hindwings.

Urea The main nitrogenous waste in mammals and aquatic vertebrates. It is very soluble in water and is therefore easily excreted as urine.

Ureter The duct which carries urine from the kidney to the bladder in mammals.

Urethra The duct from the urinary bladder to the outside of the body in mammals.

Uterus The mammalian womb where the embryo is nourished by the PLACENTA. The lining of the placenta undergoes cyclic changes during the reproductive life of the animal (see OESTROUS CYCLE and MENSTRUAL CYCLE).

Vaccine A substance injected into the bloodstream of a person or animal which protects or immunizes the recipient from attack by a specific disease. A vaccine is usually made from a weakened or dead form of the organism which causes the disease, or from the organism's toxins (poisons). It works by stimulating the production of antibodies so that if the person is ever exposed to attack by a healthy form of the organism, the body's defence mechanisms come into play immediately to prevent the disease from developing.

Vacuole A fluid-filled space within the protoplasm of a cell. Most plant cells have a large central vacuole filled with sap.

Vagina The birth canal connecting the uterus with the outside of a female mammal.

Valerian Any of about 300 species of herbs and shrubs of the northern hemisphere. Common valerian is a perennial herb with deeply cut leaves and pale pink, funnel-shaped flowers. Its stem is up to 1.5 metres tall. The flowers are fragrant, but the rootstock has a nauseating smell attractive to cats and rats. Red valerian has glossy leaves and is widely cultivated. It often grows on walls and rock-faces.

Vampire bat Bats of tropical and subtropical America that feed on the blood of birds and mammals. Contrary to popular belief, they are very small creatures — less than 10 cm long. They have razor-sharp teeth which, aided by an anti-coagulating and anaesthetic saliva, enable the bat to effect a painless wound which bleeds freely.

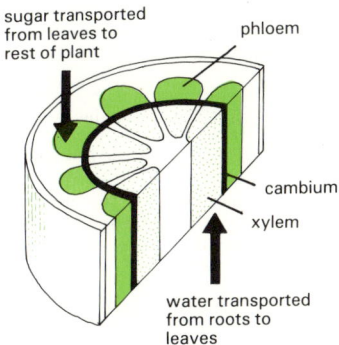

Vampire bat

The blood is simply lapped up as it wells to the surface. These bats are dangerous because they can carry the dreaded disease called rabies.

Vanessid A group of butterflies which includes some of the most colourful and best known species of the northern hemisphere, such as the RED-ADMIRAL, the PEACOCK, the PAINTED LADY, the TORTOISE-SHELLS and the CAMBERWELL BEAUTY.

Vanilla Any of several species of tropical climbing orchids. Vanilla has fleshy leaves and light green flowers. The fruit is a cylindrical pod from which the vanilla flavouring comes.

Vapourer moth A moth remarkable for the fact that the female of most species is wingless and never leaves the cocoon in which she has pupated. The male is a small active day-flying moth. Several species are serious pests of fruit trees, which the caterpillars strip in summer.

Variety An animal or group of animals that differ from the typical form in one or more features and that will continue to show these differences in succeeding generations.

Varnish tree, *see* LACQUER TREE

Vascular system The system of XYLEM and PHLOEM tubes that conduct water and food materials around the bodies of ferns and seed plants.

Vector An animal that carries disease-causing organisms and other parasites and transmits them to another (host) species, e.g. tsetse flies are vectors of the sleeping sickness parasites.

sugar transported from leaves to rest of plant

phloem

cambium

xylem

water transported from roots to leaves

Vascular system

Bryophyllum

detachable
plantlet

Marram
grass

creeping
rhizome

Vegetative reproduction

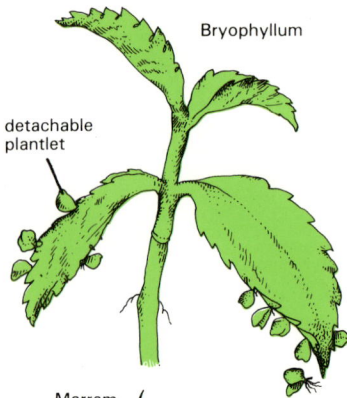

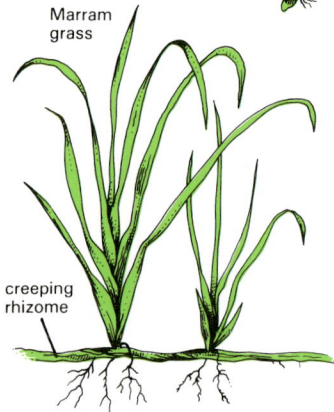

Vegetative reproduction Reproduction by the vegetative parts of a plant — root, stem and leaf — not involving the flower or any sexual process. The simplest form of vegetative reproduction is shown by many fungi and algae. The threads of which they are composed can be broken into several pieces and yet continue to grow. Higher plants, especially the flowering ones, have a number of special modifications concerned with vegetative reproduction. Many plants that creep over the ground will take root whenever their stems touch suitable soil.

Runners are specialized stems that develop on certain plants at certain times of the year and creep over the ground, taking root and developing into new plants. The strawberry is a typical example of a plant producing runners. Rhizomes are underground stems which not only spread the plant over an area but also store food and tide the plant over the winter. Corms, bulbs, and tubers are further examples of organs of vegetative reproduction. A number of plants produce detachable buds called bulbils which drop off and take root in the surrounding soil. The familiar house-plant Bryophyllum produces little plantlets on its leaves. These may often develop roots before falling from the parent.

Vein A blood vessel carrying blood back to the heart from the tissues. Also the stiffened supporting framework of an insect wing. The veins in this case carry not only blood but nerves and tracheae (breathing tubes). In plants, veins are the vascular or conducting strands of a leaf.

Velvet ant The velvet ant is a kind of wasp. It gets its name because the female is wingless and looks like a large, hairy ant. There are hundreds of species, most of them living in the warmer regions. Velvet ants lay their eggs in the nests of bees and wasps, and the young velvet ants eat the larvae and pupae of their hosts.

Vena cava, *see* HEART

Venation (1) The arrangement of the veins in an insect wing. It is important in classification. (2) The arrangement of veins in a leaf.

Ventral Concerning the underside of the body.

Ventricle, *see* HEART

Venus's flytrap A carnivorous plant native to the eastern United States. It grows in coastal swamps and is about 30 cm tall. It has long triangular leaves ending in a pair of lobes, edged with teeth, which are bright red and oozing nectar. When an insect lands on the lobes they snap shut, and do not reopen until the victim is digested.

Verbena A family of 2500 species of mainly tropical and subtropical herbs, trees, and shrubs. They include TEAK. Herbs in temperate areas include

Venus's flytrap

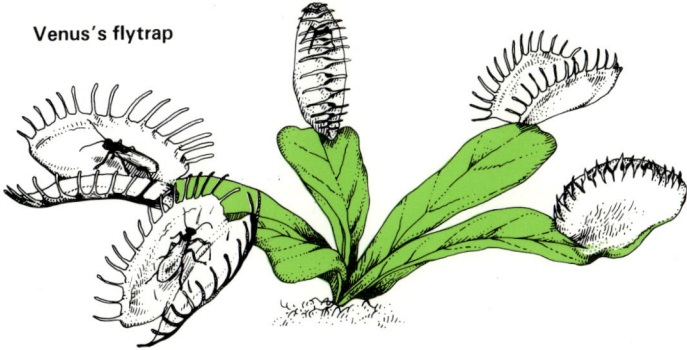

vervain, which has four-sided stems, deeply-cut leaves, and spikes of lilac flowers. It used to be called the holy herb. Several species of verbena are cultivated for their brilliant flowers.

Veronica, *see* SPEEDWELL

Vertebra One of the bones making up the spinal or vertebral column. Vertebrae are present in almost all vertebrates, replacing the embryonic notochord during development.

Vertebrate A member of the subphylum Vertebrata which includes all the backboned animals, i.e. fishes, amphibians, reptiles, birds and mammals.

Vesper bat A group of small insect-eating bats found in most parts of the world, especially in temperate latitudes. They account for about one-third of the total known species of bat.

Vestigial Of very reduced form. For example, the hindlimbs of whales have been reduced during the course of evolution until they are now represented only by tiny bones inside the body. These bones are the vestigial limbs.

Vetch Any of several scrambling plants which trail across the ground or climb up other plants by means of tendrils at the tips of their compound leaves. They are members of the pea family and have purple, blue or white flowers. Some species are grown as fodder.

Viable Able to live or develop. Used especially for embryos.

Vine The woody climbing plant whose fruits are known as grapes. There are several species and many cultivated hybrids, some of which are grown as free-standing bushes in the vineyards. The word vine is also used, rather vaguely, for any climbing plant, whether woody or herbaceous.

Violet Any of about 400 species, found world-wide. Most are small herbs, with heart-shaped leaves and five-petalled flowers with a spur at the back. The best-known species have dark bluish-purple flowers. The sweet violet is scented.

Viper The name given to a number of poisonous snakes including the true vipers of Europe, Asia and Africa and the PIT-VIPERS of America and Asia. The ADDER, often called simply a viper, and the PUFF ADDERS are examples of true vipers.

Viperfish The deep-sea viperfishes are slender pencil-like fishes up to 25 cm long. The lower jaw is relatively huge and both jaws carry long fang-like teeth. They live down to depths of 3000 metres and feed on smaller fishes and crustaceans. The viperfishes are equipped with light organs which probably attract their prey.

Viper's bugloss A biennial bristly herb of Eurasia and North Africa. It grows up to 90 cm tall with lance-shaped leaves and funnel-like blue flowers with protruding stamens. It was once regarded as an antidote to snake-bites.

Virginia creeper A climbing plant which clings to vertical surfaces with tiny sticky discs. It has five-lobed leaves, which turn bright red in autumn, and inconspicuous green flowers followed by small dark blue berries.

Two types of virus as seen through an electron microscope

a form of
food poisoning

polio virus

Virginia deer Also known as the white-tailed deer, the Virginia deer is closely related to the MULE DEER from which it can be distinguished by its habit of carrying its all-white tail raised when running. Virginia deer were almost exterminated in North America through years of slaughter by settlers. Now they are so abundant that their numbers are thought to be greater than prior to the arrival of Europeans in North America.

Virginia opossum, *see* OPOSSUM

Virus A minute organism of which many kinds are known. All produce disease in plants or animals. Viruses are so small that they can only be seen under an electron microscope. They can only reproduce within living cells. Among the many diseases caused by viruses are poliomyelitis, measles, influenza and many plant diseases known as mosaics.

Viscacha South American rodents related to the chinchilla. The plains viscacha is up to 60 cm long. The mountain viscachas are only about half that size. All live in colonies, with the plains viscacha digging extensive burrows. Mountain viscachas live among rocks and boulders.

Vitamin Any of a number of organic compounds that must be provided in an animal's diet if it is to remain healthy. Only tiny amounts are needed, but if they are not supplied certain abnormalities occur. Such abnormalities caused by vitamin deficiency are called deficiency diseases. Vitamins are generally known by a letter (e.g. vitamin A, vitamin C). Many vitamins have now been made synthetically.

Viviparous (1) Of animals: giving birth to live young. (2) Of flowers: producing vegetative buds (bulbils) in the flower clusters. The bulbils fall off and grow into new plants.

Viviparous lizard As its name implies, the viviparous or common lizard gives birth to live young. In other respects this is a typical lizard, about 14 cm long, greyish-brown above with dark spots and yellow to orange beneath with black spots, particularly in the male. The viviparous lizard is found over much of Europe and Asia, living farther north than any other reptile in these continents.

Voice The mammalian voice is produced in the larynx or voice-box by the vibration of two muscular membranes called the vocal cords. During normal breathing the vocal cords are relaxed and there is an opening between them. To produce a sound the cords are tightened by their muscles and as air is forced up from the lungs the cords vibrate like the reed of a musical instrument.

Vultures

Vole A group of small mouse-like rodents which are staple items in the diet of carnivores from owls to foxes. The field vole, as the name suggests, lives in grassy areas from pastures to gardens. The bank vole is more at home in woodlands. Voles have blunter snouts and shorter ears than mice and their teeth are also different.

Volvox A colonial green alga swimming freely in water by means of hair-like projections (flagella) around the outside. It forms a ball-like colony consisting of thousands of cells, although it is no more than 0.5 mm across. The colony may contain several daughter colonies waiting to be released, and these daughter colonies may even have their own daughter colonies inside them.

Vulpine opossum, *see* BRUSH-TAIL OPOSSUM

Vulture A large scavenging bird of prey. True vultures live only in the Old World, but the name is also applied to the condors and their relatives in America. All the vultures have more or less naked heads and most have naked necks as well. This is a useful asset because the birds plunge their heads and necks into carcasses when feeding. Vultures rarely kill and they have rather weak talons. They soar around until they find a dead or dying animal and then they swoop down for a feast.

W

Wader The name given to a large number of birds such as plovers, sandpipers and avocets which have long legs enabling them to walk in shallow water. Most also have long beaks for probing in the mud. Waders are commonly found at the seashore and in marshes, though some spend much of the time in fields and on moorland.

Wagtail Named for the way it constantly bobs its long tail up and down, the wagtail is a small bird with a total length of 15 cm. It has a fine pointed

Walrus

beak typical of insect-eaters and well developed feet with long toes. Wagtails are found over much of Europe and Asia. The yellow wagtail, the grey wagtail and the pied wagtail are three well-known species.

Wallaby Kangaroo-like marsupials living in Australia and New Guinea. Most of the species are about the size of hares, although the brush wallabies may reach lengths of one metre excluding the tail. Since the Europeans introduced foxes and other carnivores to Australia many wallabies have become rare and some have become extinct.

Wallflower A perennial European herb which often grows on cliffs and walls. It has lance-shaped leaves and, in the wild, yellow flowers. Cultivated varieties have blooms that range from cream to red and dark brown. The wallflower is a member of the cabbage family.

Walnut Any of 17 species of trees, found throughout the world. It may be up to 45 metres tall, with compound leaves containing up to 23 leaflets. The walnut bears separate male and female flowers and edible seeds contained in a woody stone and a leathery outer coat. The timber is hard and much valued.

Walrus A heavily built seal in which the upper canine teeth are enlarged to form the long tusks. It feeds on shellfish which it digs out of the seabed. Walruses live in the North Pacific and the North Atlantic. Adult bulls may weigh more than one tonne.

Wapiti A large North American deer, rather like a large red deer. It reaches a height of 1.5 metres at the shoulder. The antlers may be more than 1.5

metres long. The wapiti was once an abundant animal, found right the way across the continent, but numbers have been much reduced by hunting and the remaining animals are nearly all in reserves.

Warbler There are about 300 species of warbler, named after the melodious song of many species. These birds are generally 10–12 cm long and have fine-pointed beaks. The plumage is rather dull, with greens, browns and greys predominant. Warblers are found throughout the Old World, from Europe to Australia. The American wood warblers belong to a different group.

Warthog One of the ugliest of animals, the warthog is a relative of the domestic pig. The head is long and armed with warts and a pair of upward-curving tusks. The male is about 1.5 metres long and about 70 cm high at the shoulder. Warthogs are found in open country throughout much of Africa south of the Sahara.

Warthog

Wasp An insect belonging to the order Hymenoptera which also includes the bees and ants. There are many species, some of them social insects, others solitary forms. The common wasp of Eurasia and its North American counterparts (called yellowjackets) are familiar summer insects with yellow and black striped bodies. A queen wasp hibernates during the winter in a dry dark crevice and emerges in late spring. Almost at once she builds a small nest in a cavity — often under tree roots — from chewed wood moistened with saliva and lays the first eggs. During the summer the colony grows and the nest is enlarged by the workers to form a sphere about 20 cm across. Towards the end of the summer males and females capable of reproduction are born. After mating the males soon die and only the new queens survive the winter.

Water boa, *see* ANACONDA

Water bear, *see* TARDIGRADE

Water beetle Though less well adapted to an aquatic life than the diving beetles, water beetles live mainly in and around fresh water. The adults feed on water plants including algae and decaying vegetation. In this way they help to keep the water fresh and are consequently popular with aquarium keepers.

Water boatman A group of water bugs which spend much of their time on the bottom of ponds feeding on plant debris. They breathe from a bubble of air trapped against the abdomen. Periodically they rise to the surface to renew the air bubble. The name is sometimes applied to the backswimmer as well.

Waterbuck A large antelope always found near rivers, although it does not live in marshy ground. It is about one metre high and has a coarse brown coat. The males have long slender horns which produce a horseshoe shape when viewed from the front. Waterbuck range over much of Africa south of the Sahara.

Water buffalo, *see* INDIAN BUFFALO

Water chestnut A Eurasian and African aquatic herb. It has toothed four-sided leaves and white flowers with four petals. It produces large horned fruits which are used for food. Related species grow in China and India.

Watercress A perennial aquatic herb of Eurasia and North Africa. It has stout stems rooted in the mud and compound leaves with heart-shaped leaflets. It has white flowers which produce sausage-shaped pods. The peppery leaves are eaten as salad. Watercress is a member of the cabbage family.

Waterbuck

Water flea, *see* DAPHNIA

Waterhen, *see* MOORHEN

Water hyacinth A fast growing South American aquatic herb whose fleshy leaves often choke streams and ponds. It has violet flowers marked with red and blue. The plant has spread to rivers in most of the warmer parts of the world in the past 100 years.

Water lily Any of about 90 aquatic plants found worldwide in fresh water. They have floating round or heart-shaped leaves and large, showy flowers of white, pink, red or yellow. The royal or Victoria water lily of the Amazon has leaves up to 3.5 metres across with rims like a tray.

Water plantain Any of several species of aquatic plants, not related to true plantains. They have long or heart-shaped leaves, and small pink or white three-petalled flowers. The blooms generally grow on tall branched stems.

Water scorpion This is a flattened aquatic insect, in no way related to ordinary scorpions. The front legs are thickened and strongly hinged for grasping prey. The tail end of the insect is drawn out into a long breathing tube which can be pushed up to the surface of the water. Water scorpions live on the bottom of shallow muddy ponds and resemble dead leaves when seen from above. They eat insects and other small creatures. The water stick insect is a long slender relative of the water scorpion.

Water spider Many spiders can live temporarily under water, supported by the film of air trapped around the body, but only one species lives permanently under water. The water spider constructs a silken diving bell into which it releases bubbles of air collected at the surface. This spider ranges through temperate Eurasia.

Water strider, *see* POND SKATER

Wattle The name given to many Australian species of acacia. They have flower-heads, globular or spiked, with yellow or cream blooms. The Sydney golden wattle, growing to a height of 5 metres, is planted to stabilize sand-dunes in Europe.

Wattlebird Two rare species of bird living in New Zealand. Wattlebirds have rounded wings, long tails, and long hindclaws, with wattles, usually orange, at the corner of the mouth. They feed mainly on fruit, leaves and insects.

Waxbill A group of small colourful seed-eating birds related to the sparrows. They are confined to Africa south of the Sahara with the exception of the AVADAVAT of Asia and the Sydney waxbill of eastern Australia.

Water spider

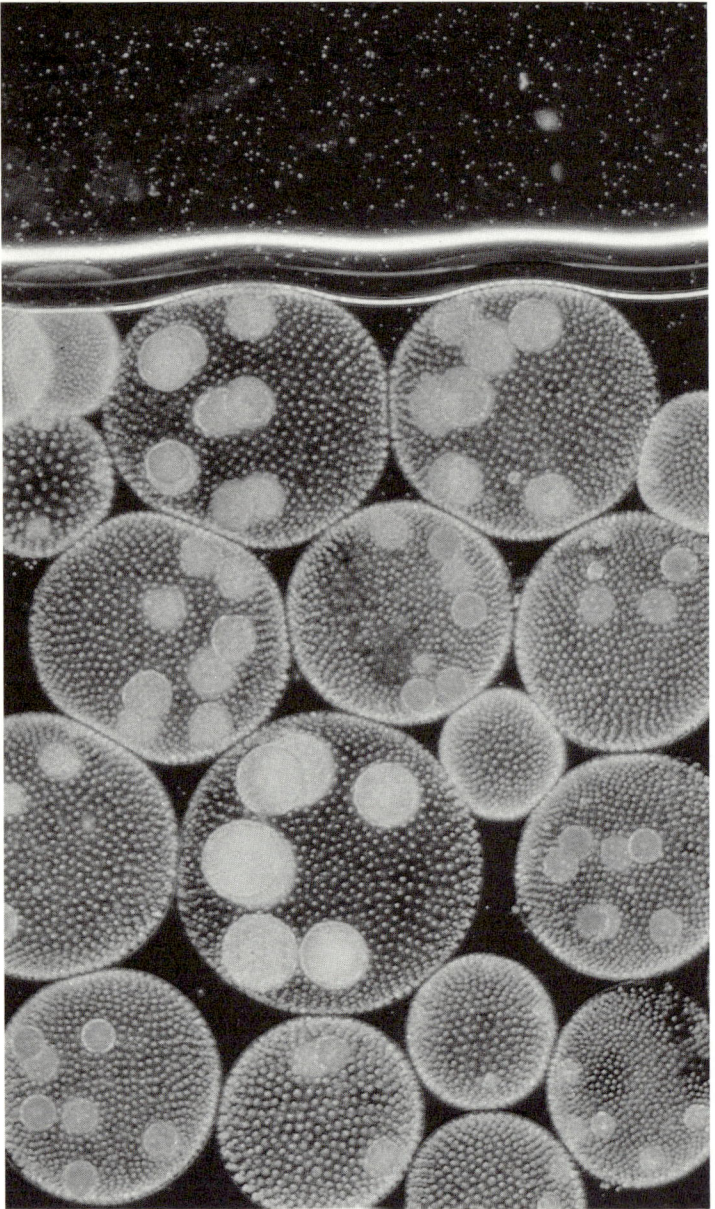

Waxwing Named after the red tips of their secondary flight feathers, which look like blobs of sealing wax, the three species of waxwing are found in the northern half of North America and northern Eurasia. Waxwings are 18 cm long, with prominent pointed crests.

Weakfish, *see* CROAKER

Weasel The weasel, found throughout Eurasia and North America, is similar in form to its near relative the stoat, although somewhat smaller, averaging about 28 cm in length including the tail. It has a slender body, short limbs and a long neck, giving it a snake-like appearance as it streaks across the ground. The fur is reddish-brown with white throat and underparts. Weasels are carnivorous animals sometimes killing animals larger than themselves, although they feed mainly on voles and mice.

Weaver bird The weavers are small, mainly seed-eating birds which live in Africa and Asia. Many of the males have bright plumage during the mating season, reverting to the drab streaky plumage of the females for the rest of the year. Weaver birds live in various types of tree country and make flask-shaped nests by weaving leaves and grasses together.

Weed A plant growing out of place or where it is not wanted. Weeds occur on waste land, roadsides and all disturbed ground, but it is in cultivated fields that they are most obvious and important. They compete with the cultivated plants for light, water and mineral salts. There are two main types of weed in cultivated land: small, quick-growing plants, often with several generations per year; and perennial herbs (e.g. bindweeds) with creeping rootstocks which continue to produce new plants even when broken into tiny pieces.

Weever A group of small fishes living in the coastal waters of the north-eastern Atlantic that are notorious for their venomous sting. Weevers spend much

Weaver bird

of the time partly buried in the sand, and a bather unfortunate enough to tread on one is liable to receive a painful sting from the poison spines of the dorsal fin and gill covers.

Weevil A group of beetles in which the head is drawn out into a snout. Most weevils are under 5 mm long. They are found all over the world and usually feed on some particular species or genus of plant, a notable example being the cotton 'boll weevil' of American cotton plantations. There are already 40,000 species of weevils known to man and several hundred new ones are being discovered every year.

Wellingtonia, *see* SEQUOIA

Whale A large aquatic mammal of the order Cetacea, whales are extremely specialized mammals. The forelimbs are paddle-like, and there is no trace externally of the hindlimbs. The tail has two large horizontal flukes which propel the body through the water. Except for a few bristles round the snout whales have no hair but, like all other mammals, they are warm-blooded and they suckle their young. There are two groups of whales: the toothed whales, and the whale-bone whales. Toothed whales (sperm whale,

A greatly enlarged photograph of volvox, a colonial alga common to freshwater ponds and measuring about 0.5mm in diameter.

porpoise, dolphin and others) have a number of peg-like teeth suited to their diet of fish and squids. The whale-bone whales, including the largest of all animals, the blue whale, feed on tiny planktonic crustaceans, which they strain from the water with plates of BALEEN that hang from the upper jaws.

Whalefish A deep-sea fish which some-what resembles a whale in appearance, though the largest are only 15 cm long. Whalefishes are black with reddish-orange patches around the mouth and fins. They are found in tropical waters at depths down to 5000 metres.

Whale shark The largest of all sharks, the whale shark grows to a length of 15 metres. The animal is dark grey, with whitish spots all over the upper sur-face. The head is broad and flattened, and the huge mouth contains hundreds of tiny teeth. Although the mouth could easily engulf a man, the whale shark is harmless. It lives in tropical seas and feeds on plankton.

Wheat A cereal grass, second only to rice in the number of people for whom it is the principal food. There are many different varieties, but they all produce nutritious grain.

Whelk The name is given to many kinds of sea snails, but the true whelk is a widely distributed species with a thick whitish or yellowish shell up to 15 cm long. It lives on the seabed below low tide level and has long been used as food. Whelk egg-cases are often washed up on the shore. They look like lumps of sponge.

Whin, *see* GORSE

Whip snake The name given to vari-ous snakes with slender bodies and long tapering tails. The dark green whip snake is the largest European snake, reaching 2 metres or more in length. It is found mainly in the Medi-terranean region. Other whip snakes are found in southern Asia, Australia and North America.

Whirligig beetle A small, shiny beetle that lives on the surface of still water. It skates round and round, searching for small insects that fall on to the water. Each eye is in two parts, one part being used for seeing on the surface, and one part for looking down into the water.

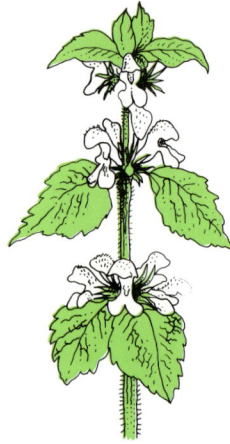

White dead nettle showing flowers arranged in whorls

Whistler A group of small insect-eating songbirds related to the fly-catchers found in Australia and South-East Asia. The golden whistler, named for the yellow breast of the male, is one of the most widely distributed species and is a common garden bird in many parts of Australia.

Whistling duck, *see* TREE DUCK

White ant, *see* TERMITE

Whitebeam A deciduous tree of Eura-sia and North Africa, whose thick leaves are densely coated with white hairs on the underside. It has flat clus-ters of white flowers and produces small red berries. It reaches a height of 25 metres.

White butterfly The cabbage white butterflies (the small white and the large white) are the most familiar European members of this group of butterflies. As the name suggests, they are pests of cultivated plants of the cab-bage family. The small white has been accidentally introduced into many other parts of the world, including North America (in the 19th century) and New Zealand and Australia (in the present century). Another European member is the Bath white whose wings are dappled with yellow and greenish-black markings.

White-footed mouse, *see* DEER MOUSE

White-headed duck, *see* STIFFTAIL
White-tailed deer, *see* VIRGINIA DEER
White whale, *see* BELUGA
Whiting An important food fish, this member of the cod family reaches a length of about 70 cm. It is easily distinguished by its silvery sides and the black spot at the base of the front fin. The whiting is found in the North Atlantic, the North Sea and the Mediterranean.
Whooper swan, *see* SWAN
Whorl A circular group of leaves or floral organs all leaving the stem at one level.
Whydah Also known as widow birds, whydahs are African seed-eating birds that lay their eggs in the nests of weaver finches. The females are drab and sparrow-like, but the males have striking breeding plumage, often with very long tail feathers.
Widow bird, *see* WHYDAH
Wild cat The European wild cat resembles a heavily built tabby cat, about 60 cm long excluding the tail which is thick, bushy and distinctly ringed. Wild cats live in the upland regions of Europe and Asia Minor. The American wild cat, or 'bob-cat', is so called because of its short stumpy tail. Grouse, rabbits and hares are the main food of these fierce carnivores.
Wildebeest, *see* GNU
Willow Any of about 300 species of trees and shrubs of the north temperate zone, with a few dwarf species in the Arctic. Willows have narrow leaves,

White willow

slender branches, and clusters of insect-pollinated catkins. They have light, durable timber. They include the OSIER.
Willowherb Any of a number of herbs with long slender leaves similar to those of WILLOW. Most have tall stems ending in sprays of rosy-purple flowers. Rosebay willowherb is one of the best known species, with 1.2-metre stems. It is sometimes called fireweed for its habit of springing up rapidly on burnt patches of ground. New Zealand willowherb, with stems 5 cm long, is a popular rockery plant.
Wilson's storm petrel, *see* PETREL
Windflower, *see* ANEMONE
Wintergreen The name of a small group of evergreen perennial herbs, native to Eurasia and North America. Wintergreens have creeping rootstocks, leathery leaves, and white, pink, or yellow flowers. The plants yield an oil used for flavouring.
Wireworm, *see* CLICK BEETLE
Wisent, *see* BISON
Wisteria Species of woody climbers of the pea family. They are native to China, Japan and North America. The Chinese wisteria is a climbing plant wth drooping flower clusters, white, pink or lilac in colour and up to 60 cm long. The compound leaves are covered with silky down when young.
Witch hazel A deciduous shrub or small tree of North America. It can

Wild cat

grow up to 7.5 metres tall. Its spiky yellow flowers appear in mid-winter when there are no leaves on the tree. A lotion is made from the bark and leaves.

Woad A biennial Eurasian herb whose greyish-green leaves yield a deep blue dye. These leaves are arrow-shaped, and the tiny yellow flowers smell like honey. They are carried in huge numbers on branching sprays. The plant grows 1.2. metres tall.

Wobbegong Also called carpet sharks, the wobbegongs have stout flattened bodies and broad heads, with a rough mottled skin. They rest on the seabed and look like seaweed-covered rocks — perfectly camouflaged to pounce on unsuspecting fishes. Wobbegongs live in the Pacific. The largest are 3 metres long but most are much smaller.

Wolf Possibly an ancestor of the domestic dog, the wolf once covered most of the northern hemisphere. Today it has disappeared from all but the wildest parts of western Europe, but it still ranges over parts of Asia and North America. Wolves stand about one metre high at the shoulder. They hunt by day, usually in small packs, and their endurance allows them to wear down all kinds of animals: wolf packs have been known to bring down a moose.

Wolfsbane A tall perennial European herb of the buttercup family, related to the monkshood and very poisonous. It has deeply divided leaves and hooded yellow flowers. Its poison was used to tip arrows and prepare bait for wolves and foxes.

Wolf spider

Wombat

Wolf spider These spiders are so-named because, instead of making webs, they run down their prey. They are smallish, drably coloured spiders. Females carry their eggs around in spherical sacs of silk which are attached to the hind end of the abdomen. Wolf spiders are very widespread. They include the famous TARANTULA of southern Europe.

Wolverine The wolverine or glutton is the largest of the weasel family, although it looks more like a badger. A fully-grown male may reach 1.25 metres in length, including about 30 cm for the tail. It has a shaggy coat of dense dark brown fur, powerful limbs, and strong claws. The wolverine lives in the northern coniferous forests and feeds on any animal it can catch. It even kills large deer.

Wombat A marsupial related to the koala. It is a stocky, short-legged creature up to 1.25 metres long and has strong claws. The two species live in southern Australia hiding in their burrows by day and coming out at night. Wombats are entirely vegetarian in their diet and feed mainly on grass.

Wood The dense plant tissue that makes up most of the stem and branches of trees and shrubs and which is surrounded by BARK. The darker HEARTWOOD is surrounded by the lighter SAPWOOD which carries sap from the roots to the leaves and the CAMBIUM layer in which all the growth of the tree or shrub occurs by cell division.

Wood ant The wood ant is noted for the large mound of pine needles it builds. This may be up to 1.5 metres in height and 3 metres in diameter. The mound is really the roof of the wood ants' nest. It keeps out the rain and helps to maintain a fairly constant temperature beneath. The nest itself consists of a large pit filled with leaf mould and earth. Beneath this are excavated chambers to which the ants retire in the winter. Wood ants are red in colour with a black abdomen. The workers are almost 1 cm in length, the queens about 1.25 cm and the males only slightly smaller.

Woodcock This is a wading bird related to the snipe. It is brown and black, about 33 cm long, with a straight bill and a short neck. There are two species: the Eurasian woodcock, which ranges from the British Isles across to Japan, and the American woodcock, which lives in the eastern half of North America. Although they are wading birds, woodcock live mainly in woodland undergrowth.

Woodchuck A large North American rodent related to the squirrels. It is a golden or reddish-brown animal, about 60 cm long, which lives in deep burrows in woods and farmlands. The woodchuck is a solitary creature and feeds mainly on grasses and other low-growing plants.

Wood duck In North America this name is given to the Carolina duck, a bird closely related to the mandarin duck and almost as brightly coloured. In Australia the name is applied to another duck, sometimes called the maned duck. Both birds nest in trees.

Wolverine

Woodlouse Woodlice are the only group of crustaceans to have successfully invaded the land. Never more than about 2 cm long, they have a small head and abdomen, but the thorax is large and makes up most of the body. It has seven segments, each with a pair of legs and a hard shield on the upper side. Some species can roll into a ball when disturbed. Although they live permanently on land, few species can withstand dry air and they live in damp places. Woodlice are scavengers of decaying vegetation.

Wood nightjar, *see* POTOO

Wood owl, *see* TAWNY OWL

Woodpecker A large group of woodland birds found in many parts of the world. They range up to 60 cm in length and are usually brightly coloured. The bill is straight and pointed. The legs are short and the tail is made up of stiff feathers which act as a support when the bird clings to a trunk. The birds have harsh calls and they also give themselves away by their

Cutaway of a tree trunk

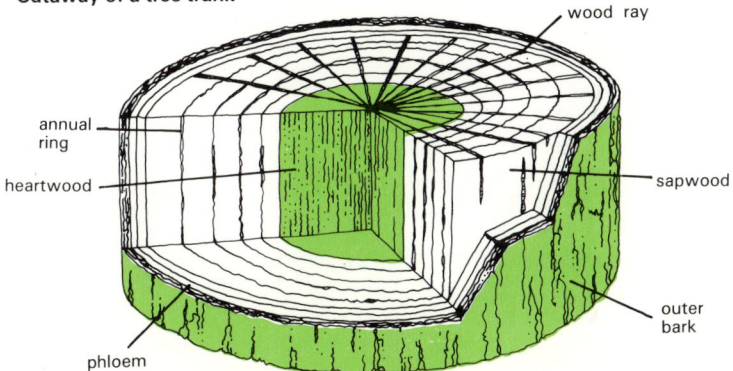

wood ray

annual ring

heartwood

sapwood

outer bark

phloem

Woodpecker

rapid hammering on the wood. Woodpeckers spend most of their time searching for insects on tree-trunks, where they also often nest, having first pecked out a hole in the trunk.

Wood pigeon Also known as the ring dove, the wood pigeon has become an agricultural pest in Europe during the last 150 years. Originally a forest bird, it has adapted to feeding in cultivated areas, particularly on grain crops. Wood pigeons are well known for their courtship behaviour of bowing and cooing to each other.

Woodruff A creeping perennial Eurasian herb. It has square stalks, around which its leaves grow in whorls of six or eight. The star-shaped flowers are white and smell like new-mown hay.

Woodrush Any of several perennial herbs related to true rushes. It is a native of Europe. It spreads by creeping runners, sending up tufts of hairy leaves with a stem up to 60 cm tall. It has clusters of pale brown star-shaped flowers.

Wood warbler A family of small, mainly insect-eating birds confined to the Americas. They range in size from 8–18 cm in length. Many have bright plumage, often yellowish in colour. Wood warblers breed from Alaska to the southern regions of South America, mainly in woodland and scrub

country. Those dwelling in northerly districts migrate to Central and South America for the winter.

Woolly bear, *see* TIGER MOTH

Woolly monkey A relative of the spider monkey, the woolly monkey lives in South America. It is about 45 cm long, with a tail of about 60 cm. The animal is covered with close, woolly fur and it has a black, rather human-looking face. The rest of the body is greyish-brown.

Worm lizard, *see* AMPHISBAENID

Wormwood Any of about 250 composite plants, mostly with strong aromatic odours. They flourish in the northern hemisphere. Common wormwood has silvery leaves and clusters of small yellow flowers, and grows up to 90 cm tall.

Wrasse A large group of marine fishes noted for their brilliant colours and their aggressiveness. Most species live in tropical waters but some, such as the cuckoo wrasse, the corkwing and the rainbow wrasse, are common in temperate waters.

Wren The true wrens are a family of birds confined to the New World with the exception of one species — the bird known in North America as the winter wren and in Europe simply as the wren. This is one of the smallest European birds, less than 10 cm long, and, like most other wrens, has a dull brown plumage. The male makes several nests and coaxes as many females as he can into them. Wrens are insect-eating birds.

Wren-tit, *see* BABBLER

Wryneck A small relative of the woodpeckers, with mottled grey-brown plumage. The name refers to the bird's habit of twisting its head right round when frightened.

Wren

XYZ

Xanthophyll One of the four major plant pigments, yellowish in colour.

Xerophyte A plant capable of survival in dry conditions or regions subject to drought. They have a number of features which enable them to do so, such as reduced leaf area, thick cuticles, and special tissues which store water.

X-ray fish This little fish from the equatorial rivers of South America is almost completely transparent. Much of its skeleton can be seen from the outside. It is an attractive fish and popular for aquaria. Although only 5 cm long, it is fiercely carnivorous and it is actually closely related to the famous piranha.

Xylem The water conducting tissues of plants.

Yak A large member of the cattle family that lives high in the mountains of Tibet. A wild yak bull may weigh three-quarters of a tonne. The animals have long, thin horns which extend sideways. They also have long shaggy black coats which reach almost down to the ground. Yaks have been domesticated as beasts of burden and also to provide milk.

Yam Any of several species of perennial climbing herbs found in most tropical areas. Yams have heart-shaped leaves, clusters of tiny flowers, and fleshy, tuberous roots, which are a major food in some countries. The name yam is often wrongly given to the SWEET POTATO.

Yarrow A perennial herb of the north temperate zone. It grows up to 45 cm tall, with feathery aromatic leaves and flat flower-heads of white, pink or purple. It is common on grassland.

Yeast Simple single-celled ascomycete fungi which normally multiply by budding. Yeasts are economically important in that they produce certain enzymes capable of producing fermentation. Yeasts convert sugars to alcohol in wine making. The baking industry uses yeast because the carbon dioxide given off during fermentation makes the dough rise. Yeasts are also important sources of vitamins.

Yellowjacket, *see* WASP

Yellow rattle A semi-parasitic annual herb with coarsely-toothed leaves belonging to the snapdragon family. It has straight spotted stems and clusters of two-lipped yellow flowers. Its seeds form in a large capsule, in which they rattle when dry.

Yellow trefoil, *see* SHAMROCK

Yew Any of a group of evergreen trees distantly related to the PINES. The leaves are flat, pointed needles. Pollen is scattered by clusters of tiny male cones. The green female flowers, borne on separate trees, give way to scarlet fruits which have harmless flesh but poisonous seeds. Yews are very long-lived, and have tough, elastic wood from which bows were made.

Yew

Yolk A food store in the eggs of most animals. It is very rich in protein. In the eggs of vertebrates (other than mammals) the yolk is found in the yolk sac which is a pouch of the embryo gut. Yolk is absorbed from the sac which gradually shrinks and is itself absorbed into the embryo before hatching.

Yucca Any of 30 species of evergreen plants of the AGAVE family. They have narrow stiff pointed leaves, and white or cream bell-shaped flowers. Some are shrubs, others are large trees. Yuccas are native to American deserts. The best-known species is the Joshua tree.

Yucca moth A small group of highly specialized moths living in Mexico and the south-western US. The larvae of the yucca moth feeds only on the seeds of the yucca plant and the plant in turn is pollinated only by the adult moth. It always produces more seeds than the larvae can eat, and so both plant and insect benefit from the association. There are even different species of moth attached to different kinds of yucca plant.

Zambezi shark A relatively common shark of warmer waters growing up to 3 metres in length. The Zambezi shark is remarkable in that ·it periodically leaves the coastal waters to swim many kilometres up estuaries and rivers. In Central America it lives permanently in the fresh water of Lake Nicaragua.

Zebra

Zebra Closely related to horses and asses, zebras are easily distinguished by their stripes. There are three species, all living in Africa. The commonest species is Burchell's zebra, in which the stripes reach under the belly. In the mountain zebra and the much larger Grevy's zebra the stripes do not extend under the belly. Zebras are the favourite prey of lions.

Zebra finch This is the most common of the Australian finches, found in most parts of the continent. The male is strikingly coloured with a black and white tail, a black-barred neck, orange cheek patches and orange legs and bill. The female is less colourful. The zebra finch is a popular aviary bird.

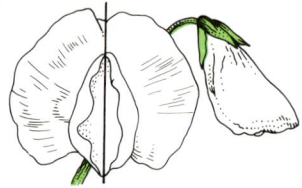

Zygomorphic flower

Zebu Docile domesticated cattle of southern Asia and Africa with a prominent hump on the shoulders and a large dewlap under the throat. The zebu or Brahman cattle readily breed with western cattle to produce hybrids.

Zinnia Any of 16 species of plants native to the southern part of North America. They have pointed oval leaves and dense composite heads composed of numerous FLORETS in a very wide range of colours.

Zooplankton The animals of the PLANKTON as opposed to the plants (phytoplankton).

Zoospore A spore able to swim by lashing tiny hairs (flagella).

Zorille The zorille looks and acts like a skunk, although it is more closely related to the weasels and polecats. It is a slender black and white animal about 60 cm long, the bushy tail accounting for nearly half the length. Like the skunk, it ejects a foul-smelling fluid when disturbed. The zorille lives in most parts of Africa.

Zucchini, *see* MARROW

Zygomorphic (of flowers) Symmetrical on one plane only; irregular (e.g. pea flower). Compare with ACTINOMORPHIC (regular).

Zygote The cell produced by the joining of two sex-cells. The zygote is the first cell of a new generation and gives rise by division to all the cells of the new individual.

The zebu is the familiar cow of southern Asia, where it is sacred to the Hindus, and of the African plains where it is the mainstay of cattle ranching.